AN INTRODUCTION TO PATHOLOGY

Campion Integrated Studies Series

AN INTRODUCTION TO PATHOLOGY

Clinical Investigation and Methods for Health Care Students

J.M. Morgan
J.A. de Fockert
C. van der Meer

Campion Press

British Library Cataloguing in Publication Data

Morgan, John
An Introduction to Pathology:
Clinical Investigation and Methods for Health Care Students
(Campion Integrated Studies Series)
I. Title II. Series
616.07

ISBN 1-873732-08-2

Campion Press Limited
384 Lanark Road
Edinburgh EH13 0LX

Cover design:
Artisan Graphics, Edinburgh

Cover photograph:
cotton wool under polarised light x 250
by J.M. Morgan

Translated sections by
E.J. Huisinga - van der Kwast

Typesetting:
Word Power, Berwickshire

Printed and bound by
Thomson Colour Printers, Glasgow

Preface

Significant advances in medicine in recent years have meant that the scope of responsibility of the health care professional has grown considerably. In the United Kingdom, as elsewhere in Europe, nursing is undergoing major changes to enable it to meet the demands of health care as we approach the 21st century. The Project 2000 nurse education programme is directed towards preparing the student for the new role of "knowledgeable doer", with practice founded on a sound knowledge base. Health care students following the new syllabus need not only a broad knowledge of the normal anatomy and physiology of the body but also an understanding of the principles of general and systemic pathology.

An Introduction to Pathology is one of the titles in the Campion Integrated Studies series. In common with other books in the series, the subject matter is presented in a clear and simple way with a clinical practice orientation. Learning objectives are clearly defined for each chapter and summaries are included to help with revision and reinforcement.

The book will be of value for Project 2000 nursing students who are following the common foundation course. It will also be useful for Trainee Medical Laboratory Scientific Officers and for students of the other professions allied to medicine.

The first part of this book, Chapters 1 to 7, considers the concepts of health and illness together with an overview of the causes of disease. This is followed by a description of general pathology topics including inflammation, infection, immunology, allergy, hypersensitivity, degenerative changes, healing repair and hypertrophy.

The final chapter in this introductory section deals with the important topic of neoplasia. It considers the difference between benign and malignant tumours and outlines their causes, classification, behaviour, diagnosis and treatment.

The second part of the book, which deals with the signs and symptoms of diseases, serves as a meaningful link between general pathology and general medicine. The importance of accurate history taking and clinical examination is stressed. The need to understand the significance of clinical findings and laboratory results in the interpretation of signs and symptoms is explained. The physiological basis of each sign and symptom is described. This provides a more comprehensive understanding of specific disease processes and enables the health care student to put individual findings into a more meaningful perspective when considering the patient as a whole. This holistic approach is in accordance with the guidelines of the current Project 2000 curricula.

Chapters 9 to 15 describe the physiological and anatomical basis of body temperature, the pulse, blood pressure, water balance, shock, oedema and respiration together with the causes and effects of various disease processes. Chapters 16 to 18 describe the types, causes and effects of anaemia, cyanosis and jaundice. This is followed in Chapters 19 to 23 by a description of the pathological basis of several important common symptoms. Chapters 24 to 29 outline the significant findings obtained from examination of urine, blood, faeces, sputum, vomitus and other body fluids and tissues.

The authors hope that this book will be regarded not merely as a catalogue list of diseases and pathological processes, but will contribute to the creation of a better understanding between health care professional and patient, to the ultimate benefit of both.

August 1993
J.M. Morgan
J.A. de Fockert
C. van der Meer

Contents

Health and ill health

1

1. Introduction

Most people take good health for granted. It is an individual's most valuable asset. Ill health is a disturbance of the normal physiological or psychological processes. This chapter demonstrates that medical science aims not only to correct these disturbances but also to try to prevent them. Pathology is the name given to the scientific study of the causes and effects of diseases. It is in effect the study of disease and disease-processes.

Learning outcomes

After studying this chapter, the student should be able to:
- explain the concepts of health and ill health;
- describe homeostasis and how it relates to health and ill health;
- identify the factors which disturb homeostasis;
- outline the different ways in which medicine is practised;
- explain the diagnosis and course of a pathological process.

2. The concepts of health and ill health

It is sometimes said that a healthy person does not need a doctor. But this is not entirely true, since the doctor's task is not only to restore health and try to relieve suffering, but also to try to prevent illness and thus preserve health.

Accordingly, a closer examination of the concept of *health* is needed. The World Health Organisation gives a definition of health as a state of complete physical, mental and social well-being. However, it must be noted that a sense of well-being is a subjective and personal feeling. A person may feel completely well and be unaware of the presence of disease in his or her body.

So it is an open question, from an objective point of view, whether a state of complete well-being can really exist. In order to be healthy the body is involved in a continuous battle, defending itself against attacks to prevent illness. This battle takes place inside very narrow limits within which efforts are made, by means of a great variety of activities, to maintain a balance between defence and attack. These activities help to preserve a reasonably stable internal physiological environment. This maintained equilibrium is called *homeostasis*, and there are many, often complicated, regulatory mechanisms which control the normal functioning of the body.

Examples of homeostatic mechanisms are: electrolyte balance, hormonal control, blood coagulation, temperature regulation, blood pressure and pulse rate control. When the homeostatic mechanisms are disturbed beyond the limits of the normal controls, illness occurs. Illness therefore is a disturbance of this homeostatic balance. The body will usually attempt to restore the balance. As a general rule, the greater the disturbance, the more the individual feels ill and the more serious the condition is.

Individual response to disease varies considerably. Age is one important factor. Babies and elderly people have poorer homeostasis and may react less well than young adults. Gender variations also exist, with females generally having more resistance to disease than males. Multiple disturbances of the balance mechanisms will obviously lead to more serious illnesses. The homeostatic mechanisms give the person resistance to disease but this resistance may be diminished by a number of factors. These can be divided into *endogenous* and *exogenous* factors.

Figure 1.1
Air pollution: an exogenous factor of disease

Endogenous factors create a disturbance or imbalance from within, and they may not always be easily recognisable. Some of them cause congenital disorders. When the disorder is transmitted from parent to offspring it is termed hereditary disease, and many such genetic disturbances have now been traced to specific chromosome abnormalities. However, many congenital disorders are not inherited and often the causative factors are not understood.

Exogenous factors (Figure 1.1) are factors which threaten resistance from the outside, such as:

- microorganisms which may cause infection;
- cigarette smoke which may cause lung cancer;
- fatty food which may cause cardiovascular disease.

The patient's mental as well as physical balance must be maintained, and the mental state may be influenced by both exogenous and endogenous factors. The disease process may thus involve the working together of both exogenous and endogenous factors and physical and mental variables. For example, physical disorder can cause the individual to become anxious or depressed. Similarly, anxiety or tension can disturb bodily function and may lead to physical disorders. Physical disorders caused by psychological influences are termed *psychosomatic* diseases. An individual's life style may have a profound effect on physical and mental balance. For example, drug dependency or malnutrition will cause serious disturbance to the balance mechanisms and may lead to permanent damage.

The old proverb 'mens sana in corpore sano' (a healthy mind in a healthy body) has been updated by Karl Barth, the Swiss theologian, to state, 'mens sana in corpore sano in societate sana' (a healthy mind in a healthy body in a healthy society) and there is certainly some justification for this view.

3. Preventive and curative medicine

Prevention is better than cure. *Preventive medicine* is that branch dealing with the prevention of disease. *Curative medicine* attempts to restore the disturbance of the balance of the regulatory mechanisms. Generally speaking prevention is only effective if the cause of a disease is known. For example, vitamin deficiency can be prevented by good nutrition in a mixed and varied diet.

4. Signs and symptoms and course of a disease

When the normal homeostatic balances are disturbed a range of changes occur which constitute the signs and symptoms of a disease. When a doctor first sees a patient, a diagnosis will be reached by a combination of observations made about the patient (*signs*), by the patient's description of what he feels or experiences (*symptoms*), and by interpretation of the patient's recollection of the course of the disease (*history*). At this stage it may not yet be possible to establish a precise diagnosis. Several probabilities may exist. This stage constitutes the *differential diagnosis*.

Although many sophisticated investigations can now be carried out, it may still be impossible to arrive at a specific diagnosis. Then it may only be possible to observe how various pathological changes subsequently develop. The course of the disease greatly helps in establishing the diagnosis. It may be either acute or chronic in character. An acute disease has a sudden onset and often develops with great intensity. When it occurs extremely suddenly and severely, the term *fulminant* is applied. One well-known fulminant disease process is lobar pneumonia. A chronic disease, on the other hand, often has an insidious onset and increases in severity over a long period of time, perhaps over several years or even a lifetime. One example of a chronic disease is rheumatoid arthritis.

Although the course and outcome of a disease, either with or without treatment, cannot be stated with certainty, the forecast of the probable course and outcome is called the *prognosis*. The prognosis of many diseases, based on the average outcome of a large number of cases, is known.

It may still be difficult to establish a prognosis for an individual patient. The doctor must try to consider all the known factors which may influence the course of the disease in that individual, but in most cases it is impossible to fully evaluate them because many will be intangible.

The course of an illness may fluctuate. After a quiet period there may be a sudden increase in the severity of the signs and/or symptoms, and this is called an *exacerbation*. The period during which there is a diminution of the signs or symptoms is called a *remission*. A *remission* may lead to an apparently complete recovery, called a *complete remission*.
Some diseases, such as urinary tract infections and duodenal ulcers, are well known for their tendency to recur.
A *complication* is the occurrence of new or additional signs and symptoms in an already existing disease. For instance, a stomach ulcer may start bleeding or perforate the stomach wall, allowing the gastric contents of the stomach to enter the abdominal cavity and cause peritonitis. A thrombus in a leg vein may detach and move to the lungs, causing a pulmonary thrombo-embolus. Children with a throat infection often acquire inflammation of the middle ear (otitis media) or an abscess in the throat.

Complications are well recognised and may seriously change the course of a disease. All health care professionals must be prepared for and recognise the occurrence of possible complications.

When the disease has been conquered there follows a recovery phase called *convalescence*. During this phase, rest and good diet are essential. Immobility due to confinement in bed will lead to loss of muscle mass and therefore weakness. Active physiotherapy may be needed and some patients will have to learn to walk

again. Irreparable or slow-to-repair tissue may, for example, result in stiff joints or unsightly scars, and even mental damage is a possibility. Convalescence in such cases will take much longer.
Rehabilitation in a specially equipped centre which offers education for independent living, to prepare patients for returning home, can be invaluable. The convalescent period for the elderly is usually far longer than for the young and often has an unsatisfactory outcome. Babies have an especially unstable (or *labile*) homeostatic mechanism and they too need particularly careful observation in the convalescent period.

Body and mind

An individual consists of both mind and body, and a disease of the body has inevitable effects on the mind. An ill person may exhibit paradoxical mental changes so that a usually phlegmatic individual may become easily irritated and emotional, and vice versa. The mental attitude of the patient considerably influences recovery, however, and so therapy must often be directed to the psyche (mind) as well as to the soma (body). Some patients have great difficulty in coming to terms with the mental effects of a physical illness.

Recovery

The concept of health is difficult to describe. It is even more difficult to assess the point at which good health is fully re-established, i.e. the moment of recovery. The doctor can only assess the measurable signs together with the subjective well-being expressed by the patient in judging recovery to be complete. The objective factors and the patient's subjective feelings do not always correspond. Sometimes measurable indices suggest that the patient has not recovered despite believing himself that he has. In other cases, although the objective factors have returned to normal the patient remains depressed or still does not feel well. Some patients will never fully recover or only recover partially, and indeed some patients will die.

Death can occur in several ways. It may be sudden and unexpected. It may follow a shorter or longer illness and be reasonably well anticipated. It may follow a long period of unconsciousness. Death has occurred when brain stem death is established, and in most cases this is not difficult to establish. When blood circulation and respiration cease, brain death follows shortly. Establishing death in a patient on artificial ventilation poses problems, and this can only be done after careful adherence to established technical guidelines. Cases where there is potential organ donation for transplantation require rigorous adherence to the procedures laid down in order to establish brain stem death.

Acceptance of death is always difficult, even when it ends suffering. Good support for both the dying patient and the relatives is of paramount importance. Aiding the patient's death *(euthanasia)* raises numerous medical, ethical, legal and religious problems, and special care is needed in a few patients suffering from hypothermia or drug overdose where signs of life can be minimal but must not be missed if a tragedy is to be avoided.

5. Summary

Health and ill health are not totally objective concepts. Individual experience to a great extent determines concepts of health and illness. This chapter has described the concepts of health and ill health, the course of illness, and death.

Causes of disease

2

1. Introduction

Aetiology is the scientific study of the cause of disease.

In humans the causes of disease can be divided into two main groups:
- internal or endogenous causes, arising within the organism;
- external or exogenous causes, originating outside the organism.

The majority of diseases are caused by exogenous factors. Diseases caused by exogenous factors may occur more easily if a latent susceptibility or predisposition to them exists.

Learning outcomes

After studying this chapter, the student should be able to:
- explain the concepts of endogenous and exogenous causes of disease;
- describe the concept of congenital disorder and give examples of various types of congenital disorder;
- give examples of several exogenous causes of disease.

2. Diseases caused by endogenous factors

This group of diseases consists of inborn or congenital disorders, meaning that the disorder has developed during the prenatal period, either anatomically or functionally as a result of an abnormal origin or development (Figure 2.1). Some of the congenital disorders result from chromosome abnormalities, others do not (Diagram 2.1).

a. Hereditary diseases

Hereditary diseases are a result of abnormality in a person's gene structure. Genes, which are located on the chromosomes within body cells, are responsible for specific anatomical and physiological properties, such as the colour of the eyes and the colour of the hair. A genetic abnormality may be transmitted from parent to child. This does not mean however that every child will inherit every abnormality. This depends on the character of the abnormality, i.e. the hereditary pattern or mode of inheritance.

It is possible to distinguish *dominant* and *recessive* hereditary diseases. In the dominant form the abnormal gene exerts a ruling influence on the healthy gene. The disorder will always be present in one of the parents. Each offspring will have a 50% chance of inheriting the disorder. Examples are hereditary spherocytosis (a disorder of red blood cells) and Huntington's

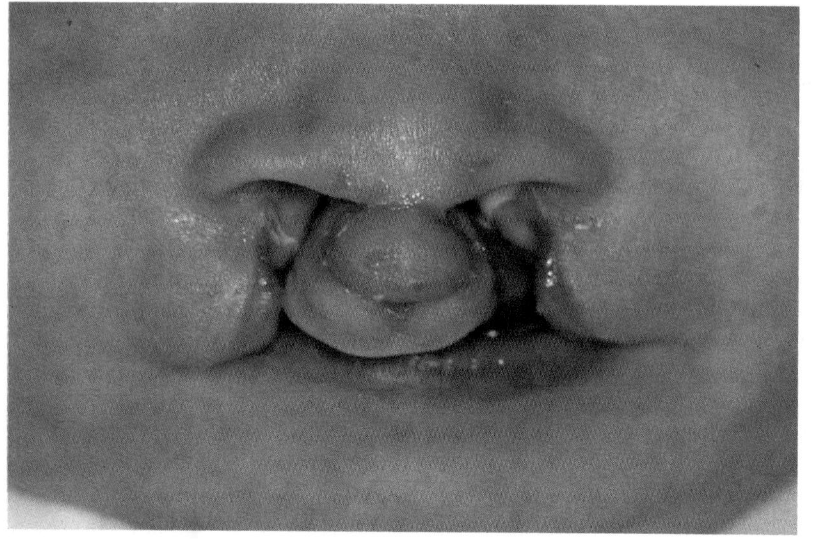

Figure 2.1
Harelip

Chorea (a form of dementia). Relatively speaking, there are only a few dominant hereditary disorders. Some can be detected by pre-natal screening and as a result some parents may decide to terminate the pregnancy; others cannot be detected by this method.

Diagram 2.1
Congenital abnormalities

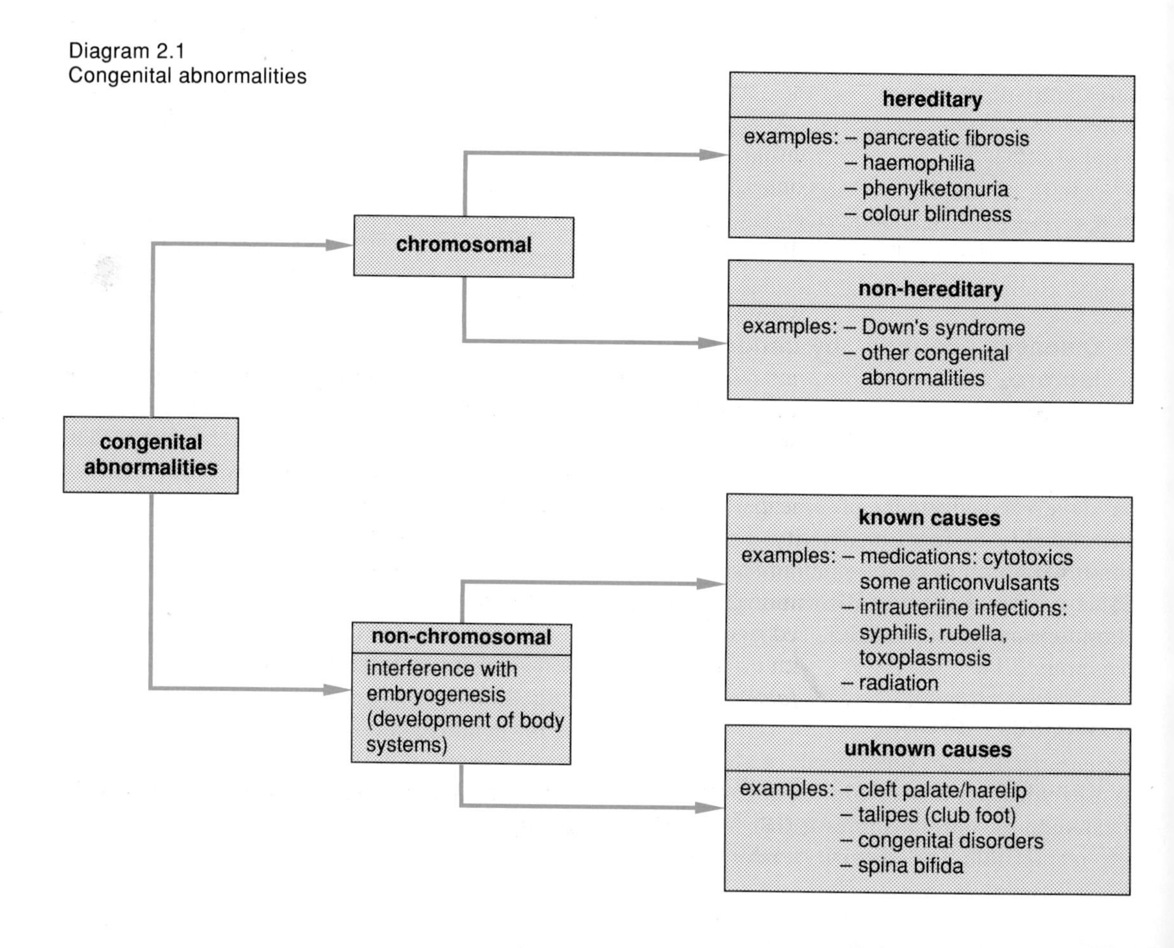

Other examples of dominant hereditary disorders include familial polyposis of the colon, and neurofibromatosis.

Recessive inheritance means that although the abnormal gene is present, it is overruled by a healthy gene. As long as the parents transmit only one unhealthy gene from one parent the disease will not be apparent in the child. The child will, however, possess the gene and be called a carrier.When both parents have the abnormal recessive gene, each offspring will have a 25% chance of inheriting the disorder. In this case both parents will be healthy although they are carriers. Examples of recessive hereditary disorders include cystic fibrosis and phenylketonuria.

Genetic counselling services are available to explain the risk factors to parents with hereditary disorders and to advise on possible outcomes.

A special variety of recessive hereditary disease is demonstrated by the condition of *haemophilia*. In this disease the genetic abnormality is located on the X sex chromosome. In females with two X chromosomes the recessive abnormal gene is controlled by the other healthy X chromosome: therefore the woman will be a carrier but will be healthy.

The male on the other hand, with only one X chromosome, will manifest the disease. Thus, as men have an XY sex chromosome make-up and women XX, females carry the gene and transmit it but do not suffer from the disease.

The frequent occurrence of some diseases is not necessarily linked with heredity. Cancer may occur in several members of a family simply by the coincidence of being a common disorder. A specific disease may be observed more often in one family than another, but with no detectable hereditary pattern. The term *familial disease* then applies.

b. Non-chromosomal congenital defects

There are, as is shown in Diagram 2.1, a large number of congenital defects which are not inherited and are not apparently chromosomally determined. The causes of these defects are often unknown. During pregnancy the mother may have contracted an infection such as rubella (German measles) or toxoplasmosis. These infections can cross the placenta and attack the fetus in utero, causing a congenital defect. In the case of rubella the baby may have heart and eye defects, in toxoplasmosis, brain disorders. Medication in pregnancy (such as Thalidomide) is a well recognised cause of limb abnormalities (phocomelia). Medication in pregnancy should therefore be avoided, especially in the first trimester (three months). Rhesus antibodies can cause haemolytic disorders in the baby when the rhesus negative mother produces antibodies to the rhesus positive blood which the fetus has inherited from its rhesus positive father.

Congenital disorders due to metabolic defects can occur.
This may lead to an accumulation of potentially toxic substances in one or more organs, and at times it may cause irreparable damage as enzyme deficiency prevents normal metabolism. X-rays or other radioactive substances can lead to developmental abnormalities in the fetus.

The causes of many non-chromosomally determined non-hereditary abnormalities are not known. Such defects include anencephaly, spina bifida, and cleft palate.

3. Diseases caused by exogenous factors

There are many exogenous causes of disease:
- mechanical causes;
- physical causes;
- chemical causes;
- malnutrition;
- biological causes;
- allergy or hypersensitivity;
- neoplasia;
- psychological causes.

Often a combination of causative factors is recognised.

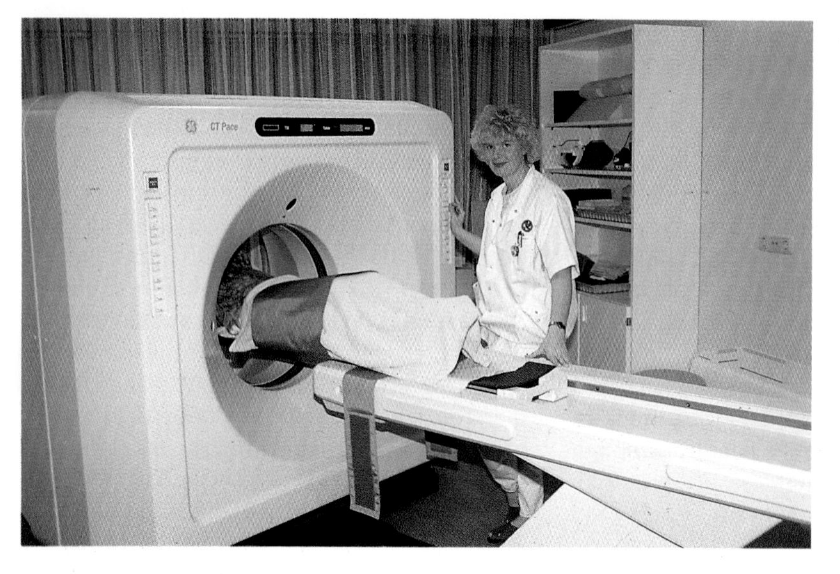

Figure 2.2a
CT scanner and radiographer

a. Mechanical causes

Direct mechanical force (trauma) can obviously cause injury. Road traffic and industrial accidents are, regrettably, all too common. They frequently result in fractures, haemorrhage, organ rupture and lacerations.
Not all traumatic injuries are acute: for example prolonged foot stress can cause so called 'march fractures'.

b. Physical causes

Physical causes of disease, apart from mechanical trauma, include:
- temperature;
- radiation;
- atmospheric pressure;
- electricity.

Temperature

Human beings are 'warm blooded', with a constant body temperature of 37 degrees Celsius. The body temperature is controlled by mechanisms such as sweating. If ambient temperature conditions exceed the body's regulatory capacity, illness will occur. An elevated whole-body temperature is known as fever or pyrexia. Local exposure to excess heat causes burning, while whole-body exposure to low temperatures results in hypothermia and local excess exposure to low temperatures can cause frostbite.

Radiation

There are several types of radiation.

Ultraviolet or UV radiation (found in the sun's rays)
Snow blindness is caused by exposure to excessive UV radiation.

X-rays
Although X-rays are used for diagnostic and therapeutic purposes, they are potentially very harmful (Figures 2.2a and 2.2b). Excessive X-radiation causes skin inflammation and can proceed to skin ulceration and even skin cancer.

Radioactive substances in small diagnostic amounts taken internally can be externally scanned, providing useful information.

Figure 2.2b
Radiation badge, worn by radiographers

Atmospheric pressure
Normal atmospheric pressure on earth is 760 mm of mercury at sea level and at 0 degrees Celsius. As altitude increases the pressure falls. At ten kilometres the atmospheric pressure is reduced to 25% of the sea level value. Above six kilometres, altitude sickness occurs.

This causes the respiration to deepen and quicken, the pulse rate to rise, and muscles to become weaker. Aircraft are pressurised to prevent altitude sickness. Many people experience popping ears in planes due to rapid pressure changes on the ear drum and the eustachian tube.

Divers are at risk from decompression sickness (called caisson disease or 'the bends'). When a diver is at depth, the extra pressure on the body causes excess oxygen and nitrogen to be dissolved in the blood. On surfacing, less oxygen and nitrogen can be held in solution in the blood and as a result gas bubbles are formed. These bubbles cause small emboli (undissolved masses) which lodge in capillaries causing severe muscular pain, breathlessness, coma, and even death. By placing the diver in a decompression chamber, pressure differences can be reduced slowly, often over several days, thus preventing the formation of gas bubbles.

Electricity
The greater the electrical current the greater the possible damage. Lightning is a natural form of electricity, and lightning strikes are often fatal. Domestic electricity at 240 volts and 13 amps is also potentially lethal. The course taken by the electrical current through the body will determine the outcome. The heart may arrest with fatal results. Yet high-frequency electricity has useful properties for therapy. Surgical *diathermy* is one example, and the use of short wave diathermy for the treatment of musculo-skeletal pain is another.

c. Chemical causes
Chemical substances may have damaging effects on the body, both locally and generally.

Caustic agents are substances which have a corrosive effect and cause damage by precipitating proteins. Examples of caustic substances include strong acids and strong alkalis. When a strong alkali such as sodium hydroxide (caustic soda) is swallowed, perhaps accidentally by a child or intentionally in a suicide attempt, the mouth and oesophagus are corroded and the oesophagus can be ulcerated or perforated. A stricture due to scar tissue may develop after healing. The eye too is very sensitive to corrosive agents. Even minor exposure can damage eyesight.

Chemical agents can cause serious systemic damage.
Parents need to be vigilant to prevent accidental poisoning in children from numerous readily available domestic products. In 1985, widespread poisoning occurred when cheap Austrian wine was adulterated with ethylene glycol as a sweetening agent. In the 1970s Spanish cooking oil was contaminated with a toxic preservative leading to numerous deaths. Addiction to drugs, alcohol or cigarettes also results in chemical poisoning.

Medicines also can act as poisons, if the dosage is excessive and suicide attempts often involve sedatives or tranquillisers. Digoxin is another example, where the therapeutic dose may be close to toxic levels.

Auto-intoxication occurs when normal metabolic products cannot be eliminated, for example in liver or renal failure. Renal failure causes uraemia, while liver failure in cirrhosis leads to hepatic coma and death.

d. Malnutrition
Health may be adversely affected by:
- a surplus of food;
- a shortage of food;
- an unbalanced diet;
- a combination of these factors.

Obesity results from a surplus of food leading to excess body fat and overweight. Too much animal fat in the diet contributes to ischaemic heart disease. Conversely, a lack of food leads to starvation, characterised by low body fat and underweight. In developing countries lack of food leads to the various problems associated wih malnutrition. These conditions are seldom seen in Western Europe; except with young women suffering from anorexia nervosa - a form of anorexia (loss of appetite) caused by psychological factors.

A balanced diet must contain *essential nutrients*, including protein, fat, carbohydrates, vitamins, minerals, and other trace elements. Calcium is needed for healthy bones and teeth. Iron is required for red blood cell production. Lack of essential nutrients causes a *deficiency disease*. Scurvy, for example, is caused by vitamin C deficiency, vitamin B-1 (aneurin) deficiency leads to beri-beri and a lack of iodine can cause thyroid disorders.

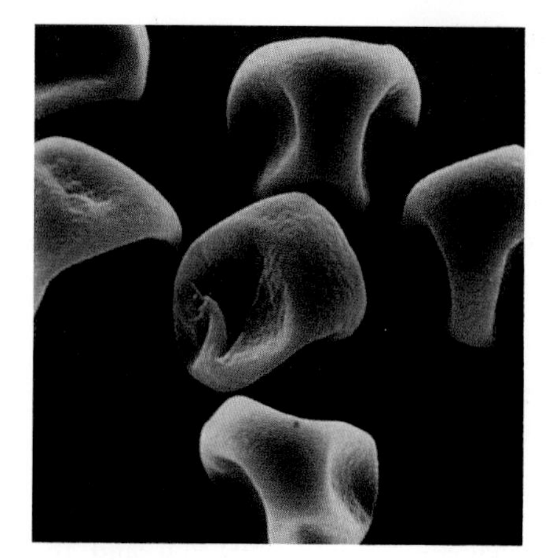

Figure 2.3
Scanning electron microscope picture of pollen grains

e. Biological causes

Diseases can be caused by infection by micro-organisms. Many different microorganisms exist, but only a few are pathogenic.

Types of microorganisms

Bacteria form the largest group. Bacterial infections may occur in almost any part of the body. Common bacterial diseases include whooping cough, tetanus, boils, gonorrhoea, and urinary tract infections.

Fungi. Many skin disorders are caused by fungi, including ringworm (which does not involve worms, despite the name), aspergillosis (which affects the lungs), and candidiasis or thrush (which may affect the mouth, oesophagus, and vagina).

Viruses cause most upper respiratory infections(including the common cold). AIDS is a viral disease.
A virus is an intracellular parasite.

f. Allergic causes

An allergy is an abnormal immunological reaction of the body to particular substances. On exposure to a foreign substance, the body may react normally by forming an *antibody*, and the

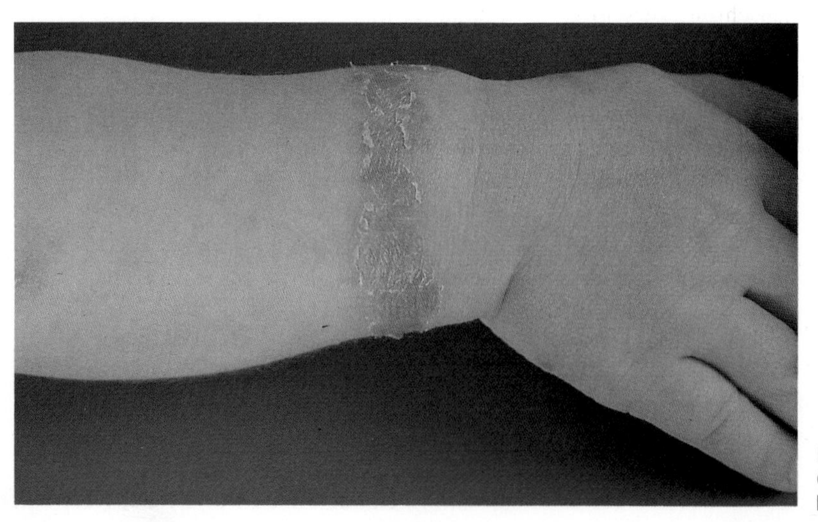

Figure 2.4
Contact dermatitis caused by nickel

substance which induces this antibody response is called an *antigen*. In normal situations the antibody production will protect the body from the antigen, but in some cases it makes the body sensitive to it. Such an allergic reaction can be fatal, for example in allergic reaction to a bee sting. Allergic reactions can be caused by medicines, food dyes, plants, fruits, metal (such as nickel), pollens, and many other substances. Grass pollens commonly cause hay fever (Figures 2.3 and 2.4).

Asthma may in some cases have an allergic component. The hypersensitivity of the bronchi to external stimuli can have an inherited component: this is called an *atopic allergy*.

g. Neoplasia

A neoplasm is an abnormal mass of tissue, the growth of which is uncoordinated with that of the normal tissues and continues in the same manner after cessation of the stimuli which have initiated it. A neoplasm may be benign or malignant. The term *tumour* is often used though strictly speaking any swelling, such as a boil, is a tumour (*tumor* in Latin means 'a swelling').

The word *cancer* is widely used to imply a malignant neoplasm. These can spread through the body by direct infiltration or through the lymphatic system or bloodstream (known as metastases).

Neoplasia can be caused by chronic irritation, such as excessive exposure to UV radiation which causes skin cancer. Cigarette smoke contains tar products which cause lung cancer. Specific viruses, *oncogenic viruses*, play a part in the causation of some cancers, and others show hereditary features. However, the reason one individual contracts a malignant disease and another does not is usually unknown.

h. Psychological causes

Psychological disorders can be caused in two ways:

- by *exogenous causes* (outside stimuli) as, for example, in the case of reactive depression following a serious life event such as a bereavement. Toxic psychoses can be caused by toxins such as *LSD* or opiates. Psychoses due to systemic illness like very high fever or liver failure (hepatic encephalopathy) also have exogenous causes;
- by *endogenous causes*, such as endogenous depression brought about without any subjective or objective exogenous cause.

Psychosomatic diseases

Psychosomatic disorders are diseases of a mental or emotional origin which have bodily signs and symptoms. Duodenal ulcers are common in people living under undue stress.
In human beings, the body (soma) and mind (psyche) constantly interact, and individuals who are ill must always be considered in their entirety, neglecting neither the psyche nor the soma.

4. Summary

This chapter has considered endogenous and exogenous causes of disease.
Endogenous causes of disease include:

- heredity;
- non-hereditary congenital abnormalities.

Exogenous causes of disease include:

- mechanical causes;
- other physical causes;
- chemical causes;
- malnutrition;
- biological causes;
- allergy;
- neoplasia;
- psychological causes.

3 Inflammation

1. Introduction

This chapter looks at inflammation, one of the most important and basic concepts in pathology. Inflammation plays a major part in many diseases, sometimes with a central role (as in pneumonia) and sometimes with a more peripheral one (as in cases of malignancy).

The causative factors of inflammation are reviewed together with the course of the inflammatory process and some aspects of treatment.

Learning outcomes

After studying this chapter, the student should be able to:
- list and describe the classic signs and symptoms of inflammation;
- identify the different causes of inflammation;
- describe the different types of inflammation and their courses;
- outline the treatment of inflammation.

2. The concept of inflammation

Inflammation is the reaction of local tissue to a harmful stimulus either from within or from outside the body.
A distinction needs to be made between *inflammation* and *infection*. Infection is just one of the many causes of inflammation.

3. Causes of inflammation

Apart from infection (invasion, multiplication, and spread of a pathogenic microorganism within the body), there are several other causes of inflammation (Diagram 3.1).

The body reacts uniformly to different inflammatory agents.

a. Mechanical causes

Tissue damage and haemorrhage can cause an inflammatory reaction without any coexisting infection. This is called a sterile inflammation, and can be seen in, for example, a mechanically caused bruise.

b. Chemical causes

Some injection fluids cause a violent inflammatory reaction if they leak from the circulatory system, although any injection can cause local tissue inflammation. Some forms of iron injections are very strongly irritant.

Contact with corrosive substances causes a local inflammatory reaction. As already described, strong acids or alkalis will, if swallowed, cause a serious inflammatory reaction to the

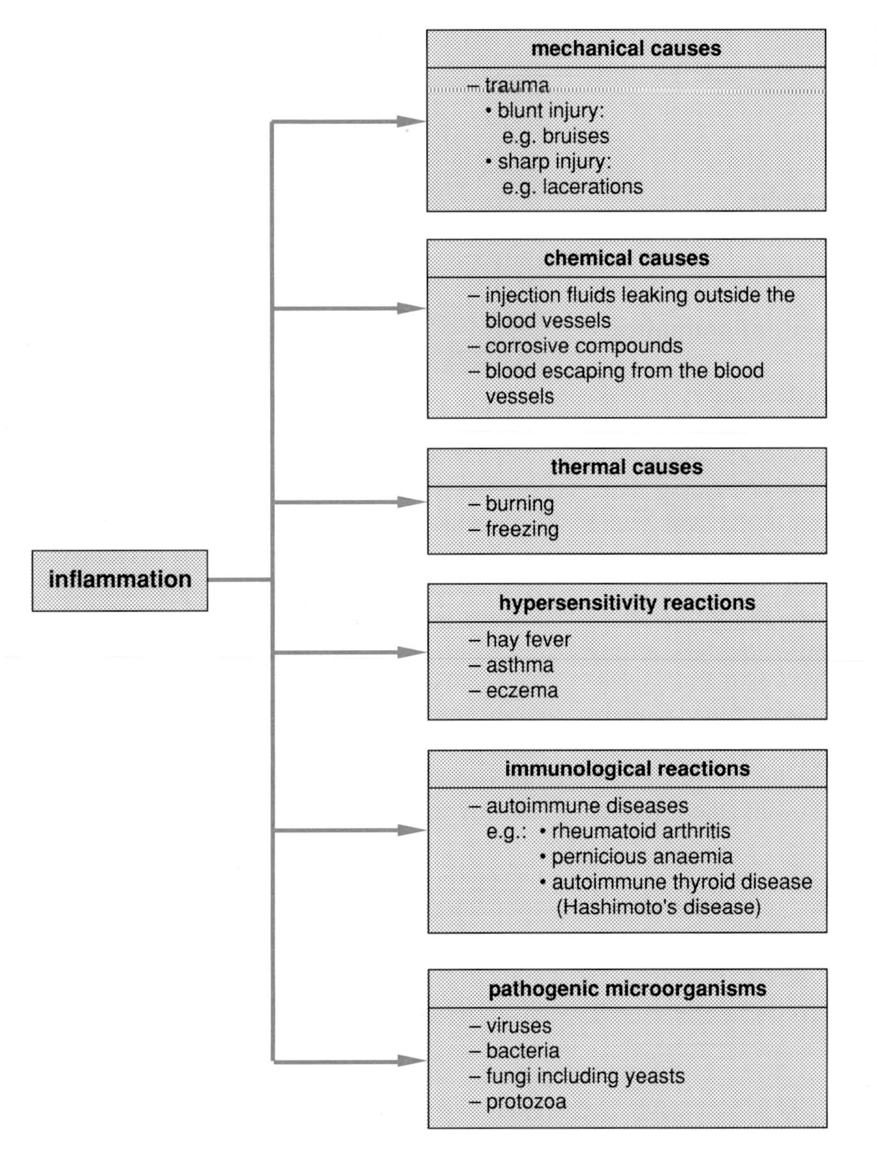

Diagram 3.1
Inflammation

oesophagus and stomach. In the case of a ruptured spleen, blood escaping into the abdominal cavity causes a chemical peritonitis. Chemical peritonitis also occurs when the intestine perforates, liberating gastro-intestinal contents.

c. Thermal causes

Both burning and freezing induce an inflammatory reaction. Sunburn is a well-known example.

d. Hypersensitivity reaction

Allergic reactions such as hay fever, asthma, and eczema cause local inflammation as a result of hypersensitivity to a foreign substance (allergen). Such disorders also have systemic effects.

e. Immunological reactions

In autoimmune diseases the body manufactures antibodies against its own tissues, resulting in inflammation. Rheumatoid arthritis causes severe and painful joint inflammation. Pernicious anaemia causes atrophic gastritis, and Hashimoto's disease causes destructive inflammation of the thyroid.

f. Pathogenic microorganisms

Commonly occurring microbiological infections are discussed in Chapter 4.

4. The clinical picture of inflammation

The early changes in an inflammatory process involve dilation of the blood vessels, especially the capillaries. This causes redness and heat.

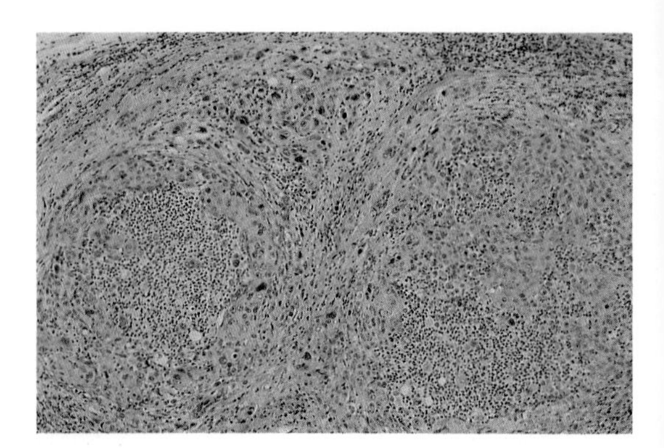

Figure 3. 1
The microscopic features of inflammation

Inflammation is easily seen in the skin, and a throbbing pain is common. As a result of vasodilation, fluid migrates from the blood vessels and the volume of extracellular fluid increases. The pressure caused by the swelling irritates the nerves and causes pain. Leucocytes (white blood cells) also leave the blood vessels. These leucocytes combat and neutralise pathogenic microorganisms by the process of phagocytosis (Figure 3.1). Inflamed tissues have impaired function.

The five classic signs of inflammation are therefore *redness* (rubor), *heat* (calor), *swelling* (tumor), *pain* (dolor), and *loss of function* (functio laesa) (Figure 3.2). The inflammatory reaction counteracts the harmful local stimulus, but the patient may nevertheless feel generally ill. Signs and symptoms may include fever, loss of appetite, and tiredness.

The course of the inflammatory process varies from individual to individual depending on site, resistance, type of inflammatory agent (the precise variety of microorganism, for example) and any treatment.

Generally, three outcomes are recognised: complete resolution of the inflammation without complication, resolution with residual complications, and failure to resolve possibly proving fatal.

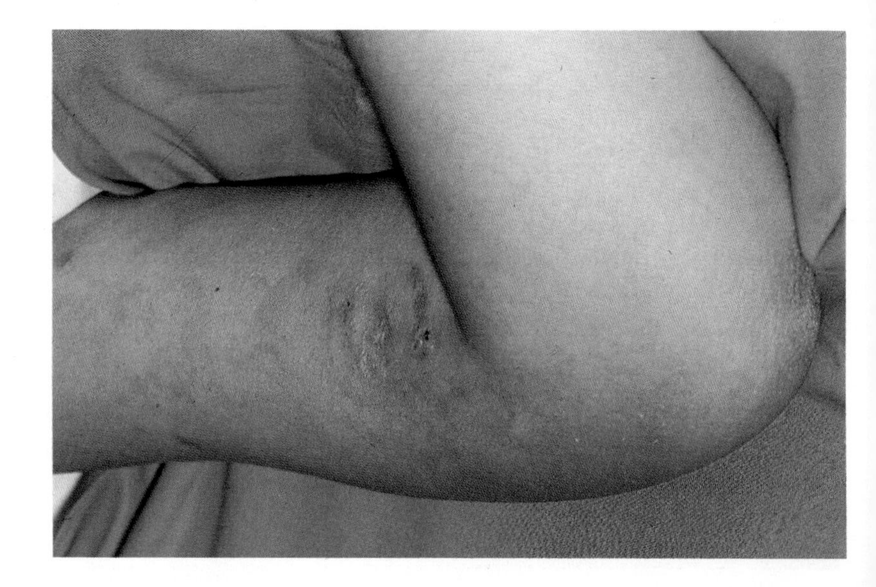

Figure 3.2
An inflamed elbow

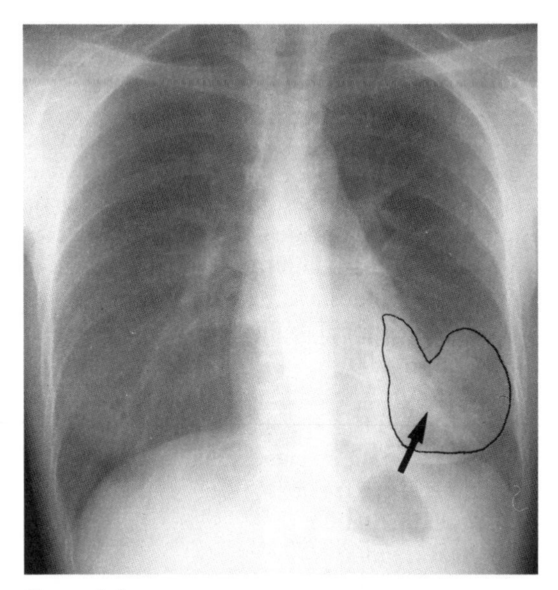

Figure 3.3
Pneumonia

Types of inflammation

Inflammation of a mucous membrane results in catarrh and is accompanied by a copious discharge of mucus. Nasal catarrh is well recognised in a head cold.

When the deeper tissues are involved the accumulating cells are referred to as an *infiltrate*. This infiltrate makes the tissues feel more solid than normal. Deep areas of inflammatory infiltrate can only be identified by special diagnostic procedures, such as the use of X-rays to identify pneumonia (Figure 3.3). If the area of inflammation is severe the tissues may die, leading to necrosis (such as occurs, for example, in a peptic *ulcer* of the stomach). In a similar way, skin ulcers on the lower leg in the elderly may be slow to heal(Figure 3.4).

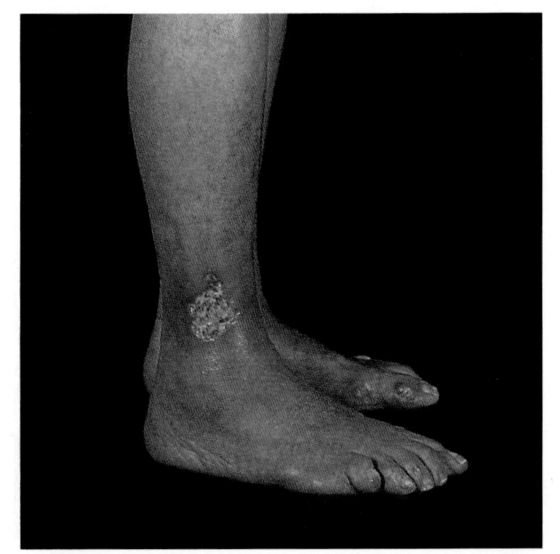

Figure 3.4
Leg ulcer

An *abscess* is a cavity formed as a result of inflammation. It contains a fluid accumulation rich in leucocytes, necrotic tissue debris, proteins and, often, microorganisms. This fluid is called pus. Pus forming in an already existing cavity is called an *empyema*, such as an empyema of the gall bladder or the pleural cavity.

5. The course of inflammation and the body's defence mechanisms

a. Course

The course of an inflammatory process depends on site, resistance, inflammatory agent, and treatment.

b. Defence

Defence mechanisms may be direct, as described here, or more indirect and immunologically mediated, as described in Chapter 5.

Capillary dilation leads to an increase in local blood supply and the arrival of numerous leucocytes. These permeate the vessel wall and infiltrate the extravascular space. New leucocytes are produced by the stimulated marrow. The leucocytes will neutralise pathogenic microorganisms by phagocytosis, and if necesary medication such as antibiotics may be used to aid the natural process.

Resolution

When the inflammatory agent is overcome, the tissues return to normal by a process of resolution. The inflammatory infiltrates subside and local healing occurs.

Abscess formation

Local tissue destruction can lead to abscess

formation. As already stated, an abscess contains pus made up of dead tissue, leucocytes, protein, and living and dead microorganisms.

An abscess leads to systemic signs and symptoms such as fever, tiredness and, usually, local pain which is often throbbing. Healing requires the pus to be drained, either by natural discharge or by surgical intervention. An abscess may break through to a blood vessel and disseminate any infectious organisms it contains, but more often the abscess cavity is walled off by local tissue reaction and contained.

When an abscess discharges to the skin surface, resolution usually follows. However, in a deep-seated organ such as the intestine, the abscess can discharge into the peritoneal cavity, resulting in potentially fatal peritonitis. A lung abscess (Figure 3.5) can break through a bronchus and pus will be expectorated.

6. Treatment

Effective treatment of an inflammation must be directed at its specific cause. Details can be found in standard textbooks of therapeutics, but a few general principles can be briefly described here.

Spontaneous healing depends on the resistance of the patient, the body's inflammatory response and the efficacy of the immunological system. Supportive symptomatic treatment such as rest, good diet and pain control will help. Specific treatment of an abscess may include antibiotic therapy and surgical drainage.

7. Summary

Inflammation can be caused by:
- blunt or sharp trauma;
- chemical irritation;
- thermal trauma;
- hypersensitivity reactions;
- immunological reactions;
- pathogenic microorganisms.

This chapter has described the clinical signs and symptoms of inflammation together with an account of the inflammatory process, and has outlined its treatment.

Figure 3.5
Lung abscess

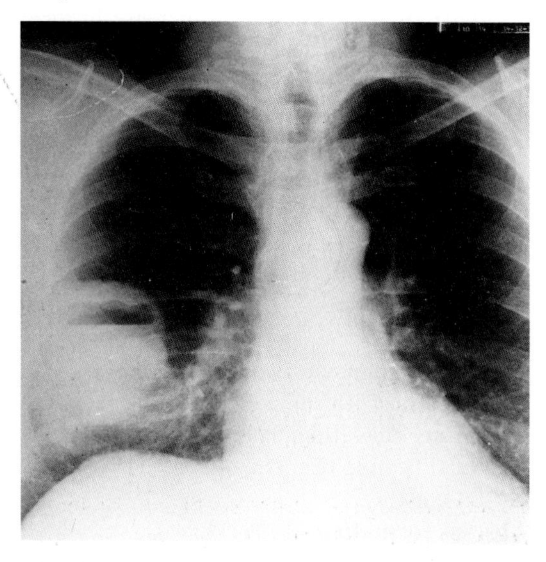

Infection

4

1. Introduction

Infectious diseases are very common. They are caused by an invasion of microorganisms. In some cases, a particular microorganism is responsible for a specific type of infection. For example, the bacterium *clostridium perfringes* causes gas gangrene and the *rabies* virus causes rabies. However, in many cases, such as wound infections, a wide variety of microorganisms can be implicated. They include *staphylococci, streptococci* and'coliforms'. Simultaneous infection by more than one type of microorganism is common.

Whenever possible, microbiological examination of appropriate specimens should be undertaken before antibiotic therapy is commenced.

Antibiotics are effective for most infectious diseases, but they are ineffective against viruses.

Learning outcomes

After studying this chapter, the student should be able to:
- identify the routes of infection;
- present a classification of pathogenic microorganisms;
- describe the course and complications of infectious diseases;
- outline the treatment of infectious diseases.

2. Concepts

a. Pathogenic microorganisms

Pathogenic means disease-producing. In medical microbiology a pathogen is a microorganism which causes infection. However, it should be noted that not all microorganisms are harmful.

A patient whose resistance to infection is reduced, for example by an underlying disease such as diabetes, or by immunosuppressive therapy to counter rejection following transplant, is particularly vulnerable to an infectious disease.
The majority of microorganisms in the environment are totally harmless. The human body harbours numerous innocuous and often useful microorganisms. These are called *commensals* or *normal flora.*

Some microorganisms behave pathogenically in certain parts of the body; for example *escherichia coli,* normally a non-pathogen when resident in the intestines, is a common pathogen if it turns up in the urinary tract.

b. Contamination

Usually, the presence of microorganisms does not result in infections, either because of low *virulence* (power to cause disease), or because the number of organisms is insufficient to cause disease.

c. Infection

Infection is the invasion and multiplication of microorganisms in the body. Inflammation (see Chapter 3) is the local reaction of tissue to a pathogenic agent and is a common feature of infection.

3. Routes of infection

Infection can occur via various routes including:
- the respiratory tract(airborne infection)
- the alimentary tract (enteral infection)
- a break in the skin (cutaneous infection)
- the bloodstream (haematogenous infection)

a. Airborne infection

Infected droplets are transmitted by sneezing and coughing. Influenza or the common cold are good examples of airborne infections: 'Coughs and sneezes spread diseases'.

b. Enteral infection

Contaminated food and drinking water may transmit microorganisms. These invade the mucous membrane and may reach other tissues via the bloodstream. Examples of common enteral infections include typhoid, cholera, poliomyelitis, hepatitis A, diarrhoea, and dysentery.

c. Cutaneous infection

Intact skin is highly resistant to infection, though some fungi (dermatophytes) can infect moist skin, in 'athlete's foot' for example. Invasive systemic fungal infections are very dangerous and difficult to treat.

Staphylococcus aureus causes boils and carbuncles. Erysipelas is caused by *streptoccus pyogenes* (Figure 4.1).

The presence of a surgical suture predisposes to local skin infection. Weil's disease (caused by a leptospiral spirochaete) breaches damaged skin or, sometimes, the conjunctiva. This organism contaminates water exposed to rats' urine. The tetanus bacterium gains entry through puncture wounds but needs an anaerobic (oxygen-free) environment to multiply.

d. Haematogenous infection

A true haematogenous infection is one where the infective agent is introduced directly into the bloodstream. This is clearly the mechanism in

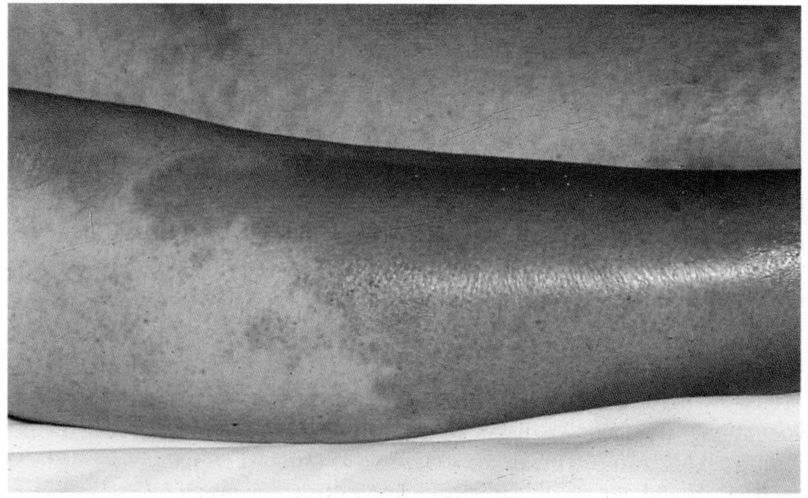

Figure 4.1
Erysipelas of the lower leg

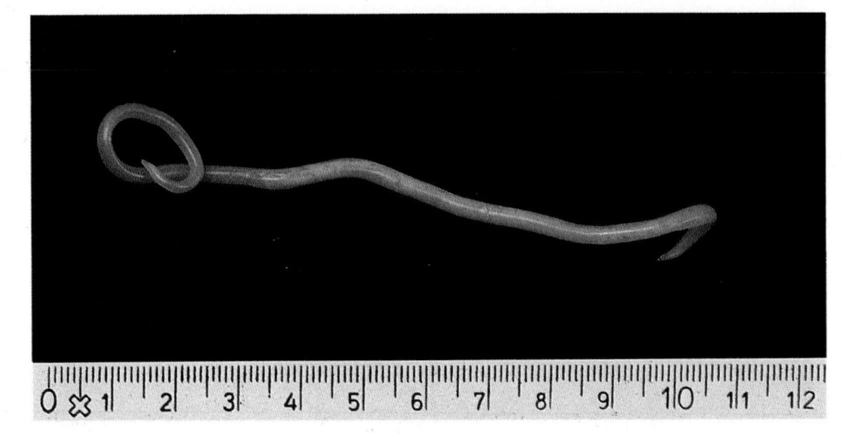

Figure 4.2
Roundworm (*Ascaris*)

malaria, where the parasite enters the bloodstream directly from a mosquito bite. Other infections, for example hepatitis B and HIV can be transmitted via a contaminated injection needle or sexual intercourse. Syphilis can be spread by haematogenous route via the placenta from mother to fetus.

4. Infectious diseases

Although by definition any microbial infection causes diseases, the term *infectious disease* is generally used to refer to a communicable disease such as mumps or chicken pox where a specific pathogen causes a specific illness and, after recovery, the resultant immunity to further attacks is usually life-long.

5. Types of pathogenic microorganisms

Protozoa

Protozoa are single-celled microorganisms. Amoebic dysentery and malaria are protozoal infections. The advent of cheap air travel leads to the importation of these normally tropical diseases to colder climates. Malaria is particularly important, since untreated it may prove fatal. *Giardia lamblia,* often acquired on a Mediterranean holiday, is responsible for a memorable if not dangerous diarrhoeal illness.

Worms and flukes

This large group of multicellular organisms is biologically diverse and especially important in tropical countries. Examples include roundworms (*Ascaris*) (Figure 4.2), hookworms (*Ancylostoma*), whipworms (*Trichuris*), threadworms (*Oxyuris*), tapeworms, liver flukes, and schistosomes. Transmission of these organisms may be direct or by an intermediate host (for example, cows are the intermediate hosts for beef tapeworms). It is worth noting that 'ringworm' is a skin infection caused by a fungus and ***not*** by a worm.

Fungi

Candida albicans (Figure 4.3) is a fungal infection, generally of a mild nature, causing for example vaginal thrush. In cases of immunosuppression the fungus can give rise to a life-threatening systemic infection.

Candida albicans can readily multiply in patients on antibiotic therapy when normal commensals are killed.
Athlete's foot is caused by dermatophytic fungi, mostly of a trivial nature but on occasion leading to more serious secondary systemic bacterial infection gaining access through the damaged skin.

Lung infection by *Aspergillus fumigatus* is a potentially fatal fungal infection. Treatment is very difficult.

Bacteria

Isolation and identification of bacteria is an important function of a microbiology department. Specimens for 'culture and sensitivity' must be collected without contamination and submitted for examination without delay.

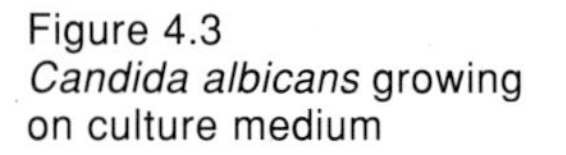

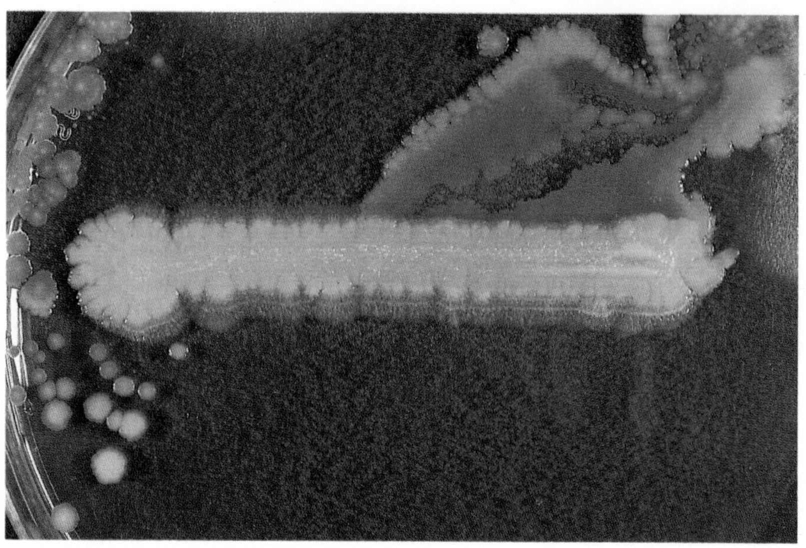

Figure 4.3
Candida albicans growing on culture medium

One important laboratory test is a *Gram's stain*. Bacteria stain either *Gram positive* (blue) or *Gram negative* (red), and this enables microscopic assessment of shape grouping (identifying, for example, streptococci or staphylococci) and staining reaction. Tubercle bacilli are a notable exception and require a different staining technique (the Ziehl-Neelsen stain) as they do not stain with the Gram method.

The culture of bacteria from, for example, a pus swab from an infected wound, takes at least 24 hours and often longer to test for antibiotic sensitivity.

In broad terms bacteria may be classified by shape:

- cocci (round);
- bacilli (rod-shaped);
- spirochaetes (corkscrew);
- vibrio (comma-shaped).

Cocci

Cocci are spherical bacteria. They include streptococci, staphylococci, pneumococci, meningococci, and gonococci.

Streptococci (Figure 4.4) occur in pairs or in chains and stain Gram positive.

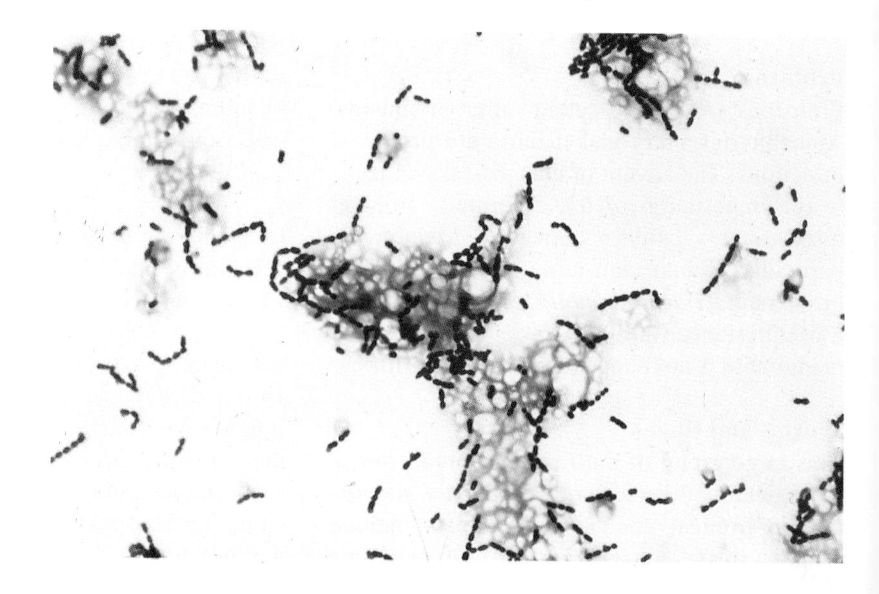

Figure 4.4
Streptococci; gram positive (1000x)

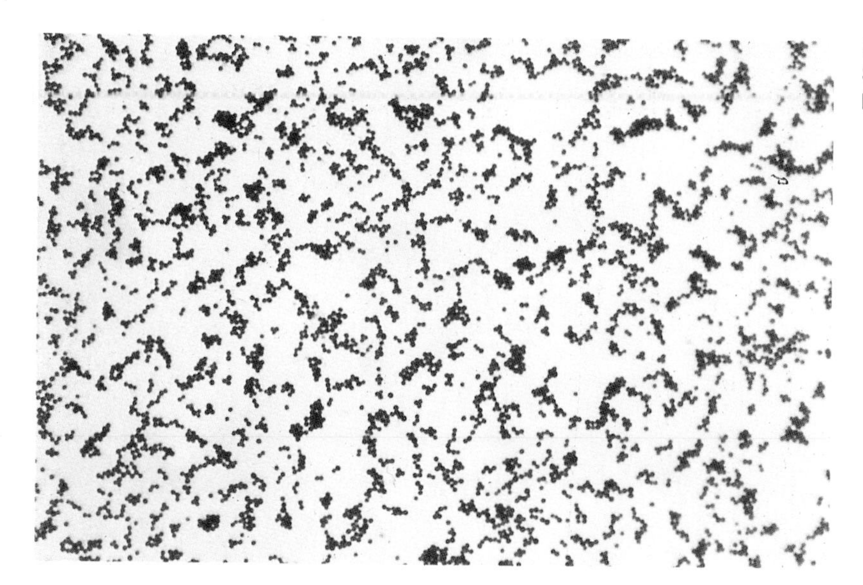

Figure 4.5
Staphylococci; gram positive(1000x)

Many varieties of *streptococci* can be identified, and some are normal commensals. Streptococcal infections commonly cause sore throats, and they usually respond to penicillin.

Staphylococci (Figure 4.5) occur in clusters and are Gram positive. They occur in suppurative infections such as boils, and they rapidly become antibiotic resistant.

Pneumococci (Figure 4.6) occur in pairs, belong to the genus of diplococci, and are Gram positive. They cause pneumonia and bronchial infections and they respond well to penicillin.

Meningococci are also diplococci but stain Gram negative. They are one of the causes of meningitis.

Gonococci are also Gram negative diplococci. They cause the sexually transmitted disease gonorrhoea.

Bacilli

Bacilli are rod-shaped bacteria, sometimes straight and sometimes slightly bent. Many types of bacilli are recognised, including tubercle, tetanus, salmonella, and pseudomonas.

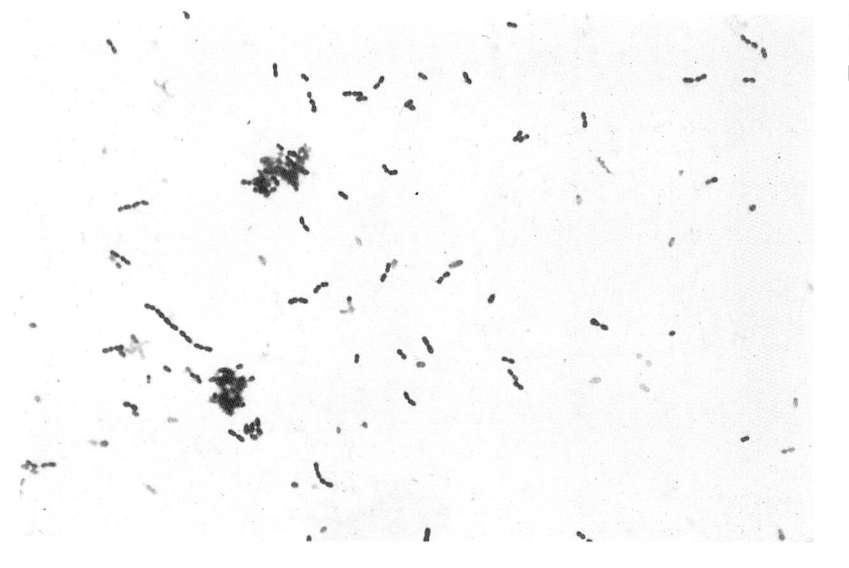

Figure 4.6
Pneumococci; gram positive(1000x)

Figure 4.7
Tubercle bacilli (1000x)

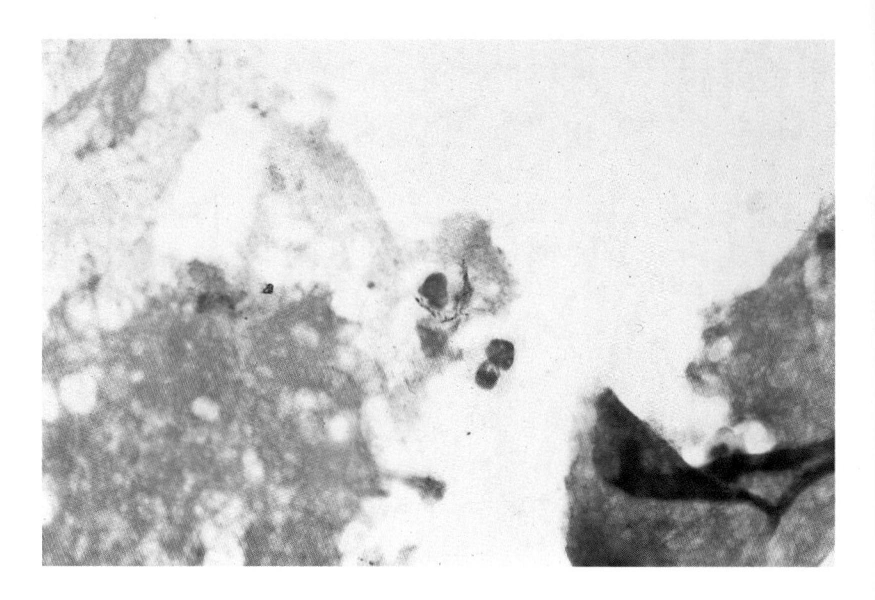

Tubercle bacilli (mycobacterium tuberculosis) (Figure 4.7), is the causative organism of tuberculosis.

Mycobacterium bovis, the bacillus causing tuberculosis in cattle, can be transmitted to humans by infected milk. Milk cattle in northern Europe are now generally free from this infection. *Mycobacterium ovis,* which affects birds, rarely infects humans.

Escherichia coli are commensals in the intestines (Figure 4.8). In this site they perform useful functions such as vitamin K manufacture. However, outside the intestines, in the urinary tract, for instance, they are pathogens.

Salmonella typhi cause typhoid.

Tetanus bacillus is rod-shaped with round ends and is an anaerobe. It is often found as spores in the soil. Infections, which usually arise from penetrating wounds, are often fatal. The organism produces a dangerous neurotoxin.

Pseudomonas and proteus occur in the intestines as commensals but are serious pathogens in other sites, such as the urinary tract. They are often antibiotic resistant.

Figure 4.8
Escherichia coli; gram negative rod (1000x)

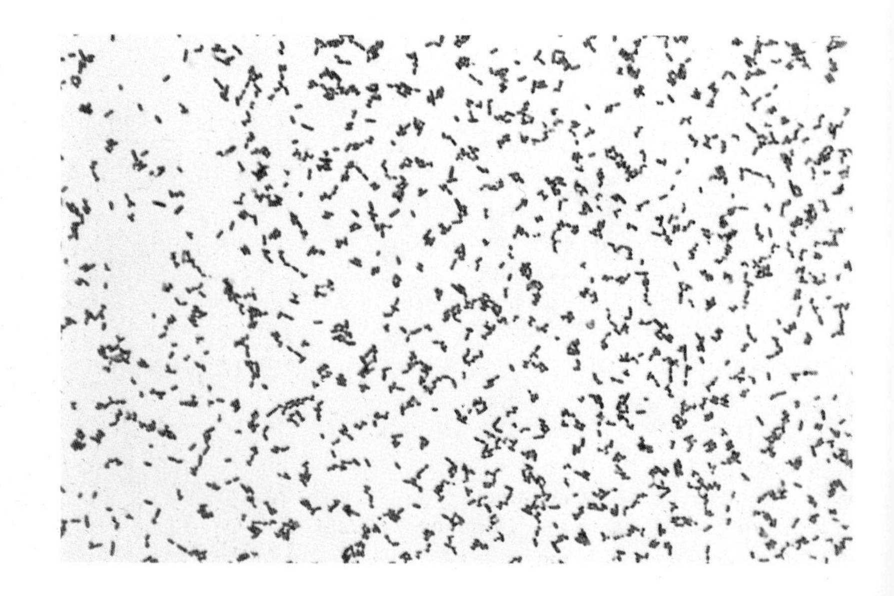

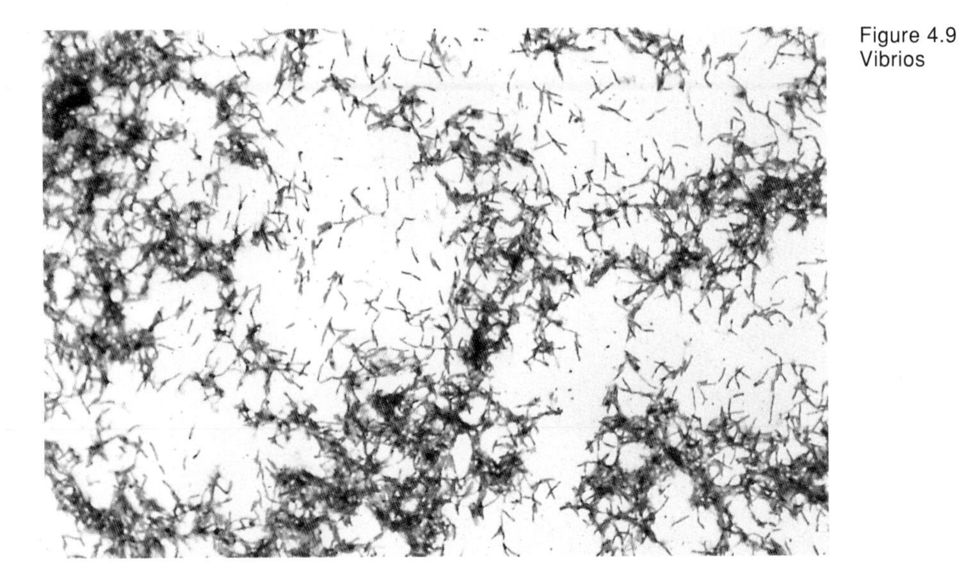

Figure 4.9
Vibrios

Haemophilus, of which several varieties are recognised. *Haemophilus influenzae* causes bronchial infections and sometimes meningitis.

Spirochaetes

Spirochaetes (Figure 4.10), are spiral-shaped. Diseases caused by spirochaetes include syphilis and Weil's disease. Syphilis is sexually transmitted, whereas Weil's disease comes from water contaminated by rats' urine, the usual source of the organism. Many spirochaetes are normal oral commensals.

Vibrios

Vibrios (Figure 4.9), are comma-shaped. Cholera is caused by the cholera vibrio.

Viruses

Viruses are intracellular parasites utilising the host's enzymatic functions(Figure 4.11). They are much smaller than bacteria and do not contain all the materials necessary for independent life and reproduction – as is the case in higher life forms such as bacteria or human body cells.

Viruses can sometimes be cultured in living cells such as hens' eggs. Viruses do not respond to antibiotics. Drugs such as acyclovir have some effect on certain virus infections such as herpes zoster in shingles. Common virus infections include common colds, measles, chicken pox, German measles, poliomyelitis, hepatitis A and B, mumps, and HIV.

6. Course and defence

The body defences will try to prevent infection, and leucocytes (white blood cells which attack foreign matter in the body) and the other components of the inflammatory response will work to resolve the infection. However, if the defences fail, the infection will cause local tissue necrosis and abscess formation will occur. More seriously, if the organisms spread round the body and multiply, then septicaemia results. Septicaemia is the condition in which organisms invade the bloodstream and actually

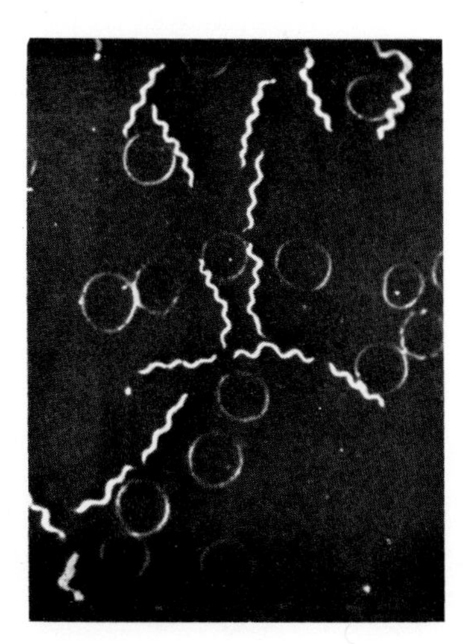

Figure 4.10
Spirochaetes

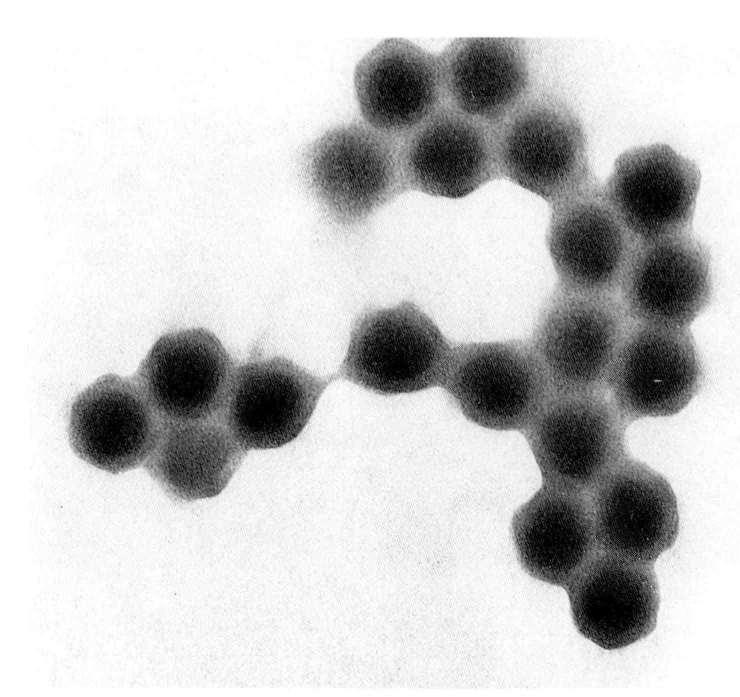

Figure 4.11
Electron-microscope picture of adenovirus (100,000x)

grow and multiply there. Sometimes these organisms are pus-forming or pyogenic and travel around in clumps or clusters which are deposited in various organs, e.g. lungs or kidneys, to form abscesses. This is known as pyaemia. Septicaemia and pyaemia are conditions often described by lay persons as blood-poisoning, conditions that are frequently fatal if untreated.

Virulence is the term used to indicate the invasive capacity of a microorganism.

7. Treatment of infections

General systemic treatment with rest, good diet and (if necessary) respiratory and cardiovascular support may suffice. Specific chemotherapeutic agents, bacteriostatics and antibiotics are often required.

a. Bacteriostatics

These agents inhibit the growth of microorganisms but do not actively kill them. Examples include the sulphonamides.

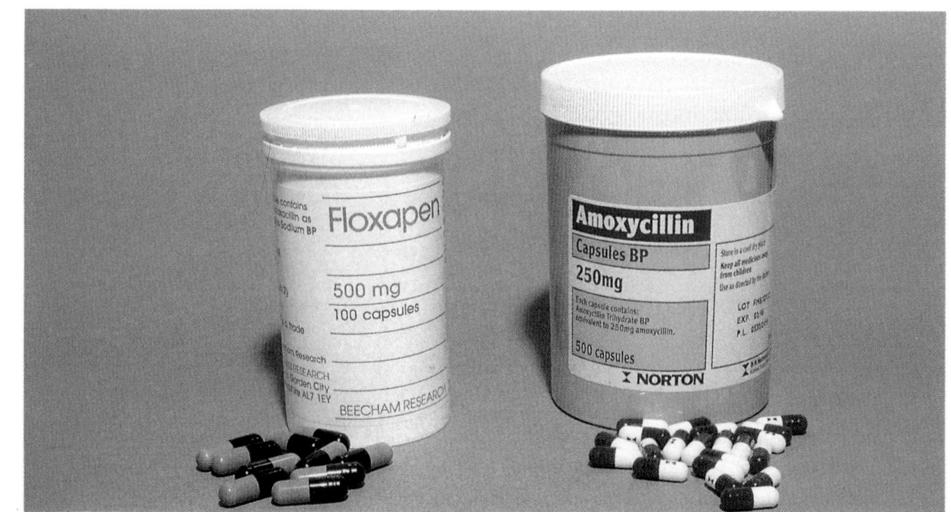

Figure 4.12
Antibiotics

b. Antibiotics

These agents are bactericidal. Many microorganisms, however, can become antibiotic resistant and pharmaceutical companies continue to develop new antibiotics to combat resistant organisms (Figure 4.12). Microbiology tests can establish antibiotic sensitivity. Broad spectrum antibiotics are effective against a wide range of microorganisms but some infections require several simultaneous antibiotics for control. Tuberculosis is a good example.

Antibiotic-resistant staphylococci are an ever present problem in hospitals. Antibiotics, like all medicines, have side effects. Normal commensals may be killed and normally harmless microorganisms can multiply. Women taking oral antibiotics, for example, frequently suffer from vaginal candida. Further details on the use and abuse of antibiotics are available in standard textbooks of therapeutics.

8. Summary

The microorganisms commonly associated with infection are:

- protozoa;
- worms and flukes;
- fungi;
- bacteria;
- viruses.

5 Immunology and allergy

1. Introduction

Immunity determines greatly why some people become ill when exposed to an infection and others do not. Allergic reactions and autoimmune disease are an expression of malfunction of the mechanisms of the immune system.

Learning outcomes

After studying this chapter, the student should have sufficient knowledge and understanding of:
- the concepts of immunology;
- the various types of immunity;
- active and passive immunity;
- autoimmunity, anaphylaxis, atopy, and tissue typing.

2. Immunology

The immune system becomes active when the body recognises the presence of a foreign substance. Such foreign substances include pathogenic microorganisms, foreign proteins, pollens, and some cosmetics.

Antigen and antibody
A foreign substance is referred to as an *antigen* if, when it invades the body, a resultant *antibody* is formed. When the body identifies a foreign protein (antigen), the immunological defence mechanism is activated. Specific types of lymphocytes (B-lymphocytes) produce protein antibodies. These antibodies are *globulins*, which can be measured to assess the immunological response. Humans possess many inborn normal globulins, and great care must be taken when blood is cross-matched to avoid serious antigen/antibody reactions.

3. Autoimmunity

Sometimes the body does not recognise its own proteins and it produces antibodies against them. This is referred to as *autoimmune disease*. Pernicious anaemia (in which gastric mucosal antibodies are produced), haemolytic anaemia, rheumatoid arthritis and autoimmune thyroiditis (Hashimoto's disease) are types of autoimmune disease.

4. Immunity

Active immunity occurs when the body produces antibodies itself. *Passive immunity* occurs when the antibody is administered. Either type of

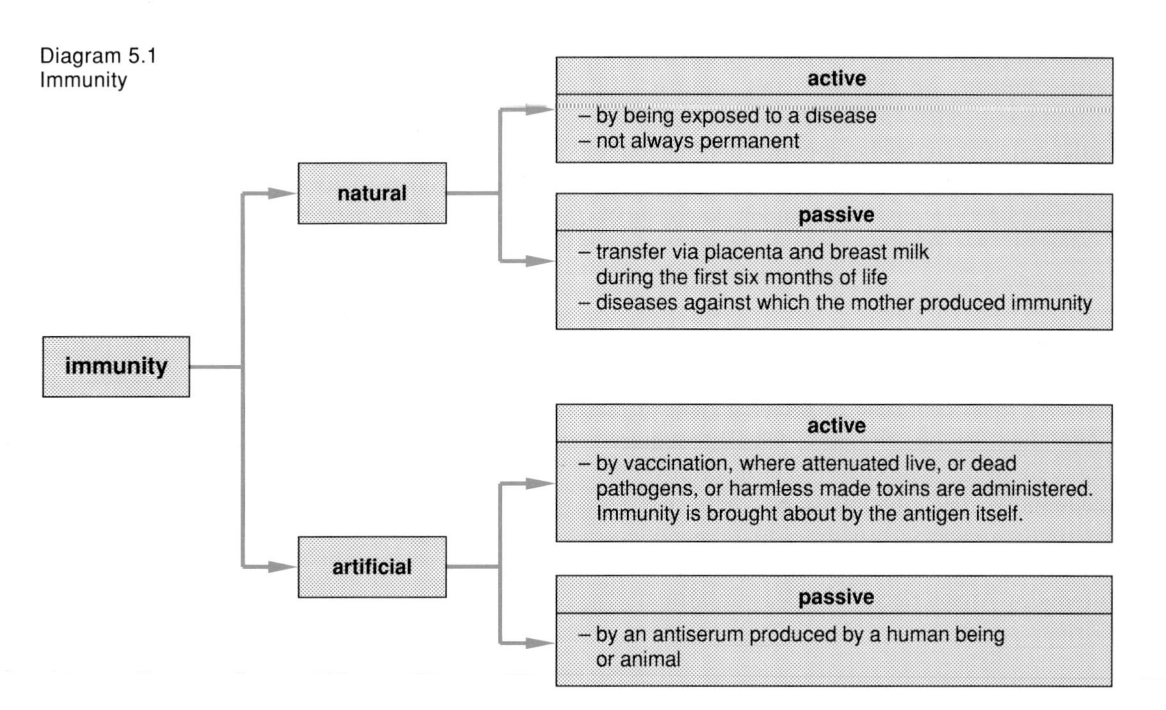

Diagram 5.1
Immunity

immunity can be either naturally or artificially acquired (Diagram 5.1).

Natural active immunity to diseases such as measles and chickenpox is usually lifelong. A baby acquires natural passive immunity by transplacental antibodies from the mother, but this immunity lasts for only a few months (Diagram 5.1). Artificial active immunity or immunisation can be achieved by administration of live or dead pathogens or specific protein products such as toxins. Passive immunity can be brought about by injection of formed antibodies as a globulin.

Immunisation is readily available against measles, German measles, polio, diphtheria, whooping cough, tuberculosis, mumps, haemophilus influenzae and tetanus (Figure 5.1). People travelling to tropical countries or health care staff may need immunisation against many other infectious diseases such as typhoid, cholera, hepatitis A, hepatitis B, and yellow fever.

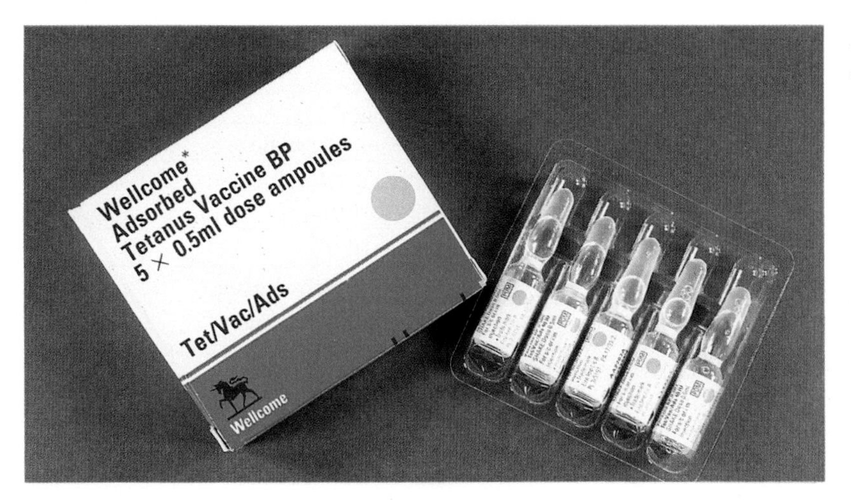

Figure 5.1
Tetanus toxoid with tetanus antiserum

Antisera produced in the serum of a third party, other human, or animal, carry the risk of introducing other foreign proteins, and the balance between the therapeutic benefit and the risk of harmful immunological reaction needs careful clinical assessment. Anaphylactic shock occurs if the reaction is sudden and severe (Diagram 5.2).

5. Allergy and hypersensitivity

Allergy is an abnormal hypersensitivity of the individual to an allergen, usually acquired through prior exposure. Such reactions may commonly be prompted by exposure to antibiotics (especially penicillin), plant pollens, foods, cosmetics, and sundry household products. Anaphylactic shock is a particularly serious variety.

Atopy

Atopy is the term used to describe allergy with an inherited predisposition. Conditions such as antibiotic allergy, asthma, and eczema often have an atopic pattern.

6. Transplantation

Transplanted organs contain foreign proteins so the normal immunological mechanisms can cause *rejection*. This can be minimised if the transplanted tissues are matched closely by tissue typing. Antigen/antibody reaction can be further suppressed by the use of immunosuppressive drugs, but these also reduce normal defence mechanisms and render the patient sensitive to infection and vulnerable to subsequent neoplasm formation.

Much has still to be discovered about immunological mechanisms.

7. Summary

This chapter has outlined:

- immunology;
- immunity;
- active and passive immunity;
- autoimmunity, anaphylaxis, atopy and tissue typing.

Diagram 5.2
Antiserum and vaccine

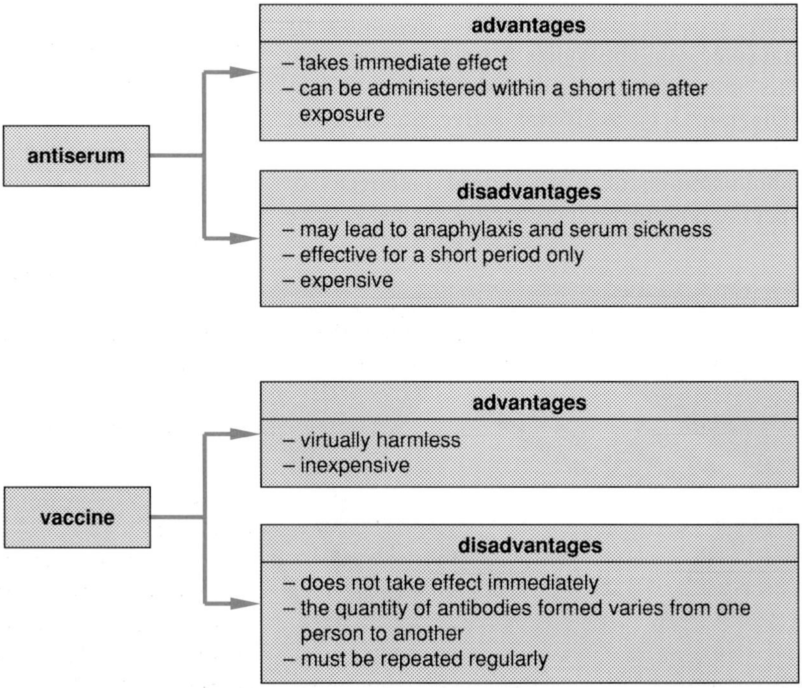

antiserum

advantages
- takes immediate effect
- can be administered within a short time after exposure

disadvantages
- may lead to anaphylaxis and serum sickness
- effective for a short period only
- expensive

vaccine

advantages
- virtually harmless
- inexpensive

disadvantages
- does not take effect immediately
- the quantity of antibodies formed varies from one person to another
- must be repeated regularly

Degenerative changes and growth alteration

6

1. Introduction

Cells and tissues can undergo all kinds of changes under the influence of external or internal factors or both. On the one hand the number of cells may decrease and their size and function may change. On the other hand the cells and organs may grow abnormally in size, and the cells increase in number, under a variety of stimuli, resulting in hypertrophy and hyperplasia.

Learning outcomes

After studying this chapter, the student should be able to:
- describe and explain degenerative changes;
- describe alteration in cell growth, hypertrophy and hyperplasia.

2. Degenerative changes

Degenerative change is examined in this chapter under the headings of:
- atrophy;
- necrosis;
- other degenerative changes and depositions.

a. Atrophy

Atrophy is the gradual decrease in size of an organ, with a reduction in the size and number of its components, and consequently a reduction in its function (Figure 6.1). Homeostasis usually maintains a balance of tissue growth.

Sometimes atrophy is a normal physiological process, as happens with the thymus after birth, and many tissues atrophy with age, such as the uterus after menopause. Local pressure can also be a cause, as in the local skin atrophy which results from bedsores. Disturbance to the blood supply or nerve supply, as happens in poliomyelitis, can lead to muscle atrophy.

b. Necrosis

Necrosis is the death of cells or groups of cells while they still form part of the living body. It may be caused by:
- a marked impairment in blood supply;
- toxins;
- immunological injury;
- infection;
- chemical poisons;
- physical damage.

Necrosis is seen in tissue infarction due to impaired blood supply. In bone infection (osteomyelitis) tissue necrosis causes the separation of dead bone, forming a sequestrum (Figure 6.2). Normally, necrotic tissue will be absorbed in due course and replaced by connective tissue.

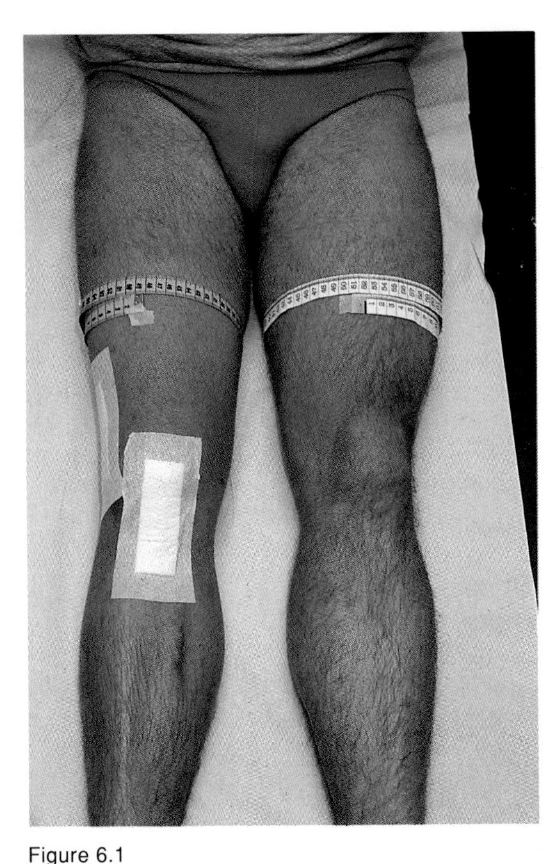

Figure 6.1
Atrophy of the muscles of the right leg after immobilisation in a plaster cast. The right thigh measures 48 cm circumference, while the left measures 52 cm

c. Other degenerative changes

These are a mixed group of tissue changes often characterised by intracellular accumulations of various compounds. They include amyloidosis, hyaline degeneration, fibrinoid degeneration, and mucinous degeneration. Pigments are deposited in tissues in many abnormal states. These include melanin, blood pigments, calcium, and lipofuscin (age pigment).

Amyloidosis

This consists of an extracellular deposition of an abnormal protein, amyloid. Amyloid is insoluble in tissues once deposited and therefore persists. The deposits may eventually interfere with the exchange of solutes between the blood in circulation and the tissue cells. Amyloid is a fibrous protein with characteristic fibrils which can be seen under the electron microscope. It is deposited outside the cells and first appears in the walls of small vessels, both veins and arteries. Even small amounts of amyloid have adverse effects on the kidneys, damaging the glomeruli and leading to proteinuria. The small vessels involved become mechanically weak and liable to haemorrhage under minor trauma, causing skin petechiae.

Amyloid has serious effects on the heart, liver, kidneys and intestines but can involve almost any organ or tissue. Some forms of amyloid are genetically determined. More commonly, occurrences are seen as a complication of chronic infection such as tuberculosis. Amyloid is also often found in cases of chronic immunological inflammatory states, the commonest being rheumatoid arthritis.

Figure 6.2
Osteomyelitis of the humerus

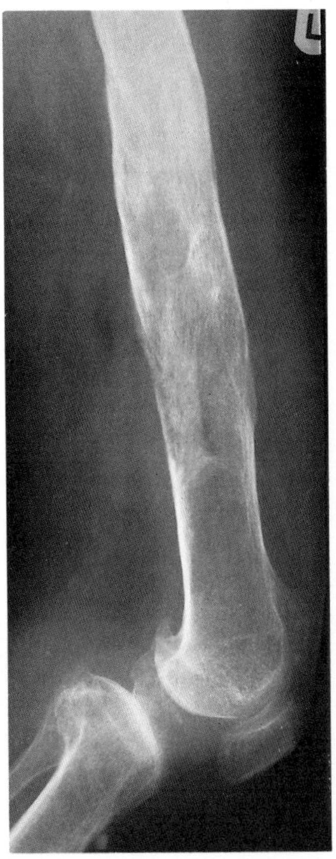

Localised nodules of amyloid are, not infrequently, found in the larynx and bladder. It is also found in the senile plaques in sufferers of Alzheimer's disease.

Fibrinoid Change

This is the release of proteinaceous material and it occurs in some inflammatory reactions, leading to fibrinoid necrosis. Examples of this may be seen in the kidneys, in the arteriolar lesions of malignant hypertension.

Mucinous and myxomatous change

Connective tissues may produce mucinous substances as a degenerative or as a neoplastic process. The term *myxoid* refers to the resemblance of these substances to the connective tissues of the fetal umbilical cord.

Melanin deposition

Abnormal melanin deposition is found in a variety of conditions. Examples include Chloasma in pregnancy, Addison's disease, vitiligo (patchy skin depigmentation) and some tumours (melanomas).

Blood pigment deposition

Haemosiderin, a breakdown product of haemoglobin, forms deposits in tissue macrophages after haemorrhage. Other abnormal iron pigment depositions may be seen in conditions such as haemochromatosis, which arises as a result of excessive absorption of dietary iron from birth onwards. The total quantity of iron in the body gradually increases and eventually may result in severe liver damage.

Lipofuscin pigmentation

As age increases, a fine brownish pigment appears in tissue cells. It is the product, mainly, of fat breakdown products. In the heart the accumulation may be marked and the myocardium eventually atrophies and becomes visibly brown (brown atrophy).

Pathological calcification

Insoluble calcium compounds can be deposited in the connective tissues following local trauma or inflammation (dystrophic calcification).

Metastatic calcification is the term applied when the deposition occurs as part of a systemic abnormality involving calcium salts. This is seen in conditions where there is excessive absorption of calcium from the intestines, such as a Vitamin D overdose.

It is also found where there is excessive mobilisation of calcium from the bones, in, for example, cases of hyperparathyroidism.

Deposition of uric acid and urates

The best known example of uric acid and urate deposition is gout, in which crystals of monosodium urate are deposited in and around joints and in nodules under the skin. The condition is often familial but sporadic cases also occur. The cause is an enzyme defect resulting in an increased breakdown of hypoxanthine into uric acid.

3. Alteration in cellular growth

An increase in the size or number of cells is seen in:

- hypertrophy;
- hyperplasia;
- repair;
- neoplasia (discussed as a separate topic in Chapter 7).

a. Hypertrophy

Hypertrophy is the increase in size of a tissue or organ without an increase in cell numbers (Figure 6.3). Exercise is well recognised as resulting in hypertrophy of muscles.

b. Hyperplasia

In hyperplasia the cell numbers increase. For instance, after blood loss an increase in haemopoesis leads to an increase in the numbers of red blood cells and leucocytes.

c. Repair

Damaged tissue is repaired by the growth of normal cells. Specialised cells may not always be restored and their place may be filled by less specialised fibrous connective tissue such as scar tissue.

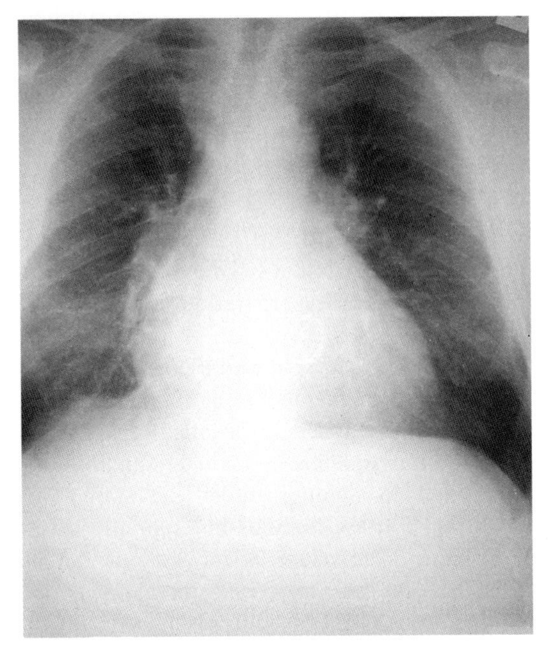

Figure 6.3
Enlarged heart -
hypertrophy

4. Summary

This chapter has outlined:
- degenerative changes;
- alteration in growth.

Neoplasia

7

1. Introduction

A neoplasm is an abnormal mass of tissue, the growth of which exceeds and is uncoordinated with that of the normal tissues and continues in the same manner after cessation of the stimuli which have started it. Neoplasia are either benign or malignant.

In developed countries malignant tumours account for approximately 20% of deaths, and only ischaemic heart disease kills more people. Cancer is feared by many people, especially if they have experienced the slow, painful and lingering death of a close friend or relative. On the more positive side, the majority of cancer suffers are elderly. Although present in all ages the incidence of cancer starts to increase rapidly after the age of thirty.

The prognosis for any individual cancer sufferer is almost impossible to predict, and depends largely on the development of metastatic spread. The outcome, however, is determined not only by the characteristics of the neoplastic cells but also by the patient's reaction to them. An understanding of the tumour/host immunological reaction may provide a basis for eventual cancer immunotherapy. The word cancer is applied loosely to all malignant neoplasms. Malignant epithelial tumours are called carcinomas while malignant soft tissue tumours are called sarcomas.

Learning outcomes

After studying this chapter, the student should have sufficient knowledge and understanding of:
- the main concepts of neoplasia;
- benign tumours;
- malignant tumours;
- causes of neoplasia;
- diagnosis and treatment of neoplasia.

Diagram 7.1
Characteristics of neoplasms

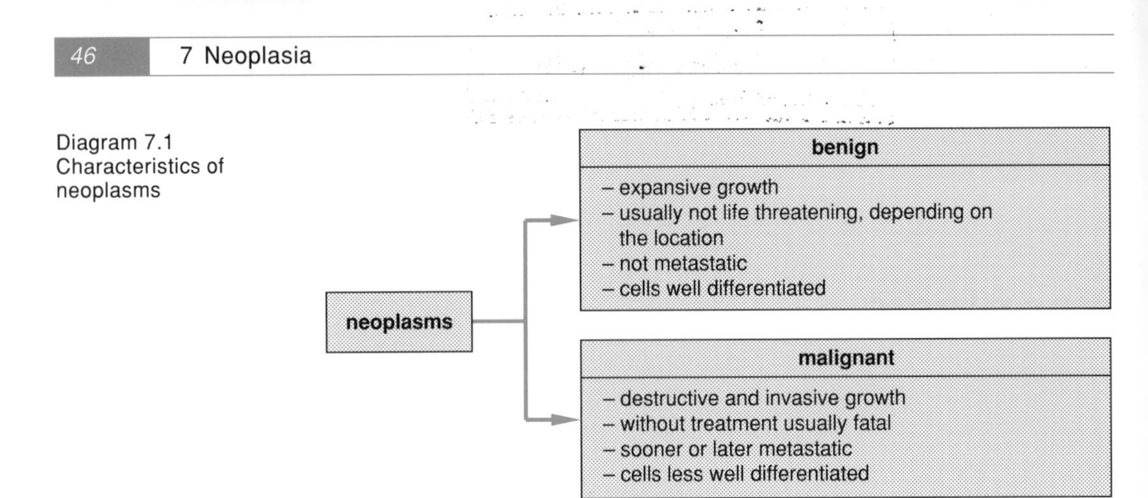

2. Classification of neoplasia

All neoplasia can be classified as benign or malignant.

a. Benign tumours

A *benign neoplasm* shows little evidence of rapid growth. It resembles the tissue of origin (it is well differentiated) and shows little potential to transgress normal boundaries. Benign tumours are non-invasive and grow locally by expansion. They are often surrounded by a compressed rim of fibrous connective tissue, forming a capsule. Although usually not aggressive, the location of a benign neoplasm in the brain, for example, may result in serious and even potentially fatal results through local pressure.

Benign neoplasia are named after the tissue of origin with the suffix *-oma*. Examples include:

– adenoma, from glandular epithelium;
– fibroma, from fibrous connective tissue;
– lipoma, from fatty tissue;
– myoma, from muscle tissue;
– angioma, from blood vessels.

Malignant change

Some benign tumours, such as adenomas in the intestine, have a malignant potential.

b. Malignant tumours

A *malignant tumour,* unlike a benign lesion, shows evidence of rapid growth, has less resemblance to the parent tissue, and tends to transgress normal boundaries (Figure 7.1). Malignant tumours spreading to distant sites are said to metastasise (Figures 7.1 and 7.2).

Figure 7.1
Uterine tumour (longitudinal cross-section showing interior and exterior of uterus)

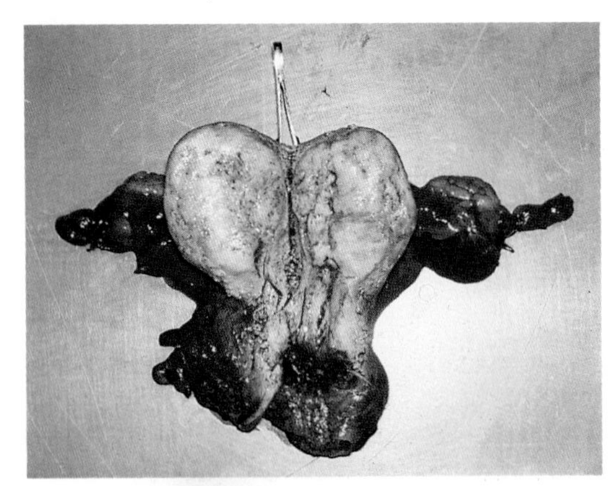

Figure 7.2
Organ metastases

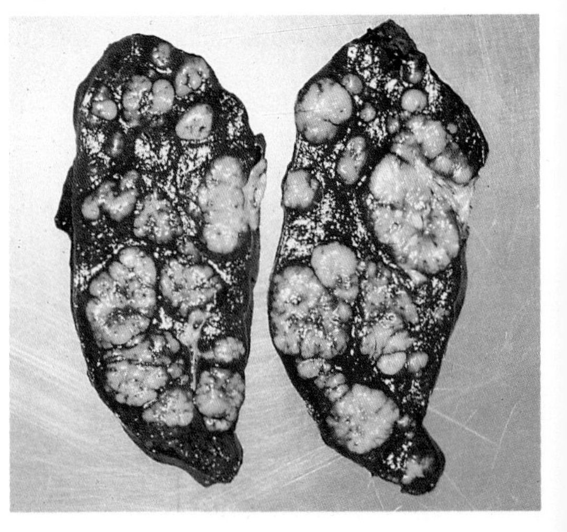

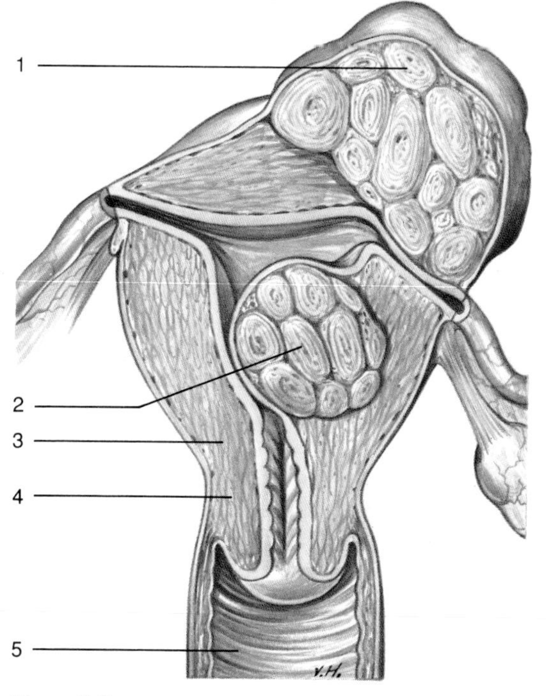

Figure 7.3
Leiomyomata in the uterus (fibroids)

1. subserosal position *2. submucosal position*
3. uterus *4. cervix*
5. vagina

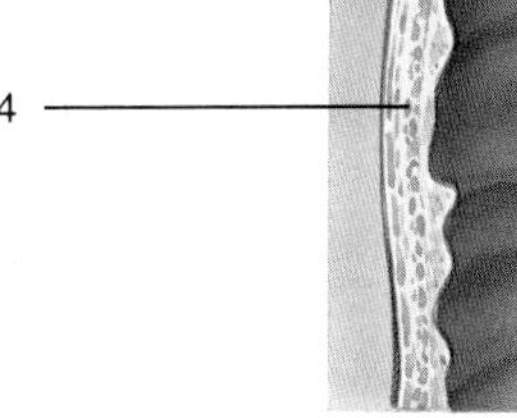

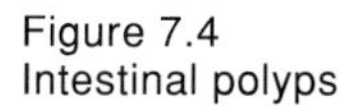

Figure 7.4
Intestinal polyps

1. small intestine *2. polyps*
3. polyp *4. intestinal wall*

Malignant tumours spread by a variety of routes:
- local infiltration;
- lymphatic spread;
- spread by the blood circulation;
- body cavity spread.

Malignant neoplasia derived from epithelial tissues carry the suffix *-carcinoma*, as in adenocarcinoma, derived from glandular epithelium. Connective tissue malignant tumours are called *-sarcomas*, as in fibrosarcoma, from the fibrous connective tissue. All tumours can thus be seen to be classified by their tissue of origin (this is called histogenesis).

Malignant tumours in a clinical situation are either *primary* (the tumour is located at site of origin) or *secondary* (the tumour has spread, or metastasised, to a remote site). A malignant bronchial carcinoma may, for example, spread from its primary site in the bronchus to form a metastasis in the liver, the secondary site.

Tumour grading is a histological estimate of the differentiation of the neoplastic cells. Tumours are graded as well differentiated, moderately differentiated, or poorly differentiated. Lesions showing no differentiation are called anaplastic tumours. The assessment is by nature somewhat subjective and open to observer variation, but quantitative morphometric techniques are being developed to overcome this variation.

Staging of tumours estimates the degree of spread of a malignant neoplasm. The tumour may be confined to the site of origin, show limited local spread, have lymph node metastastic spread or have distant metastases. The great majority of malignant neoplasia are derived from epithelium. Carcinomas account for approximately 90% of all cancer deaths.

Carcinomas may arise in any epithelial site although some sites are commoner than others. Carcinoma of the breast accounts for approximately 20% of all female cancer deaths but is rare in men. Bronchial carcinoma causes about 40% of cancer deaths in men and 15% in women. Colorectal cancers lead to about 15% of all cancer deaths in both men and women but are slightly more prevalent in women. Stomach cancer deaths, which are decreasing, cause 7% of cancer deaths in both men and women, but are slightly more common in men. Approximately 7% of cancer deaths in men are the result of prostatic carcinoma, and cervical cancer kills approximately 3000 women in the UK every year.

Hormone producing and hormone dependent tumours

Some tumours produce hormones consistent with their tissue of origin. This is called *appropriate* production. Others produce hormones not associated with their tissue of origin, and this is described as *inappropriate* hormone production. One example of appropriate hormone production is the thyroxine production from some thyroid tumours.

Some tumours, such as bronchial carcinomas, may on rare occasions produce adrenal hormones. This is an example of inappropriate production.

Prostatic carcinomas are to some extent dependent on male sex hormone for growth. Treatment may thus be aimed at reduction of male sex hormone and may include orchidectomy.

3. Causes

Many causes of individual types of cancer are known, such as the fact that cigarette smoking causes lung cancer. Nevertheless, the cause of the neoplastic transformation is as yet unknown. The process of conversion of a normal cell to a malignant one is called *carcinogenesis.*

The responsible agents are referred to as *carcinogens.*

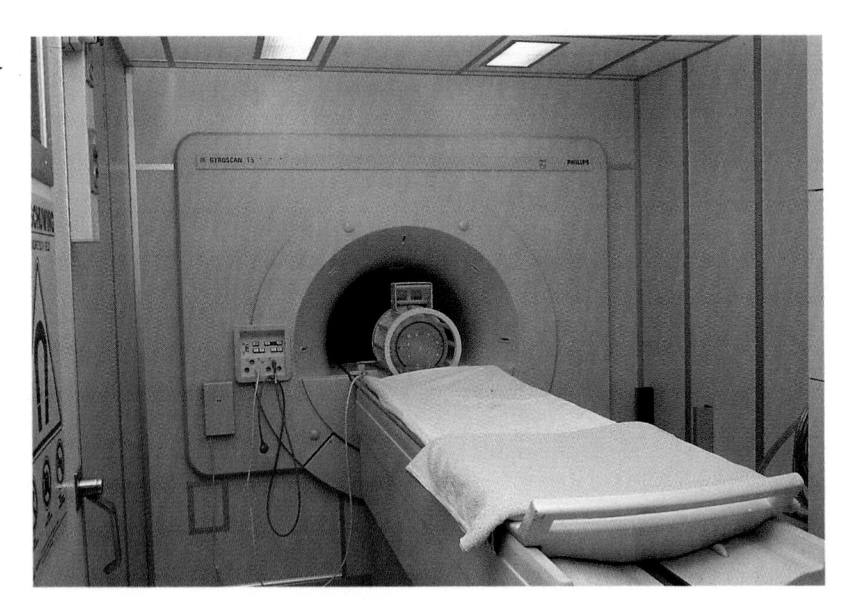

Figure 7.5a
MRI scanner

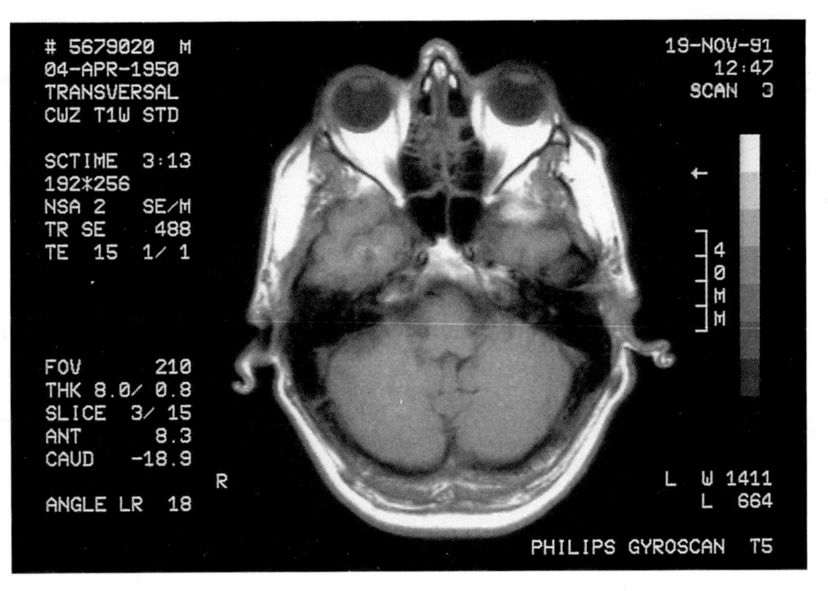

Figure 7.5b
MRI scan of head

Causal factors can be classified as:
- genetically determined;
- exogenous factors.

Genetically determined factors

Examples of genetically determined conditions include familial polyposis, which results almost inevitably in adenocarcinoma of the colon at an early age, and retinoblastoma, a highly malignant tumour of the eye in childhood.

Exogenous factors

These may include chemical carcinogen exposure, radiation, dietary factors, occupational risk, virus infection, hormones and chronic trauma.

Epstein Barr virus is involved in certain types of lymphoma (Burkitt's), aniline dye workers have a high incidence of bladder cancer and UV exposure causes skin cancers. On a global basis, hepatitis B virus is the commonest aetiological

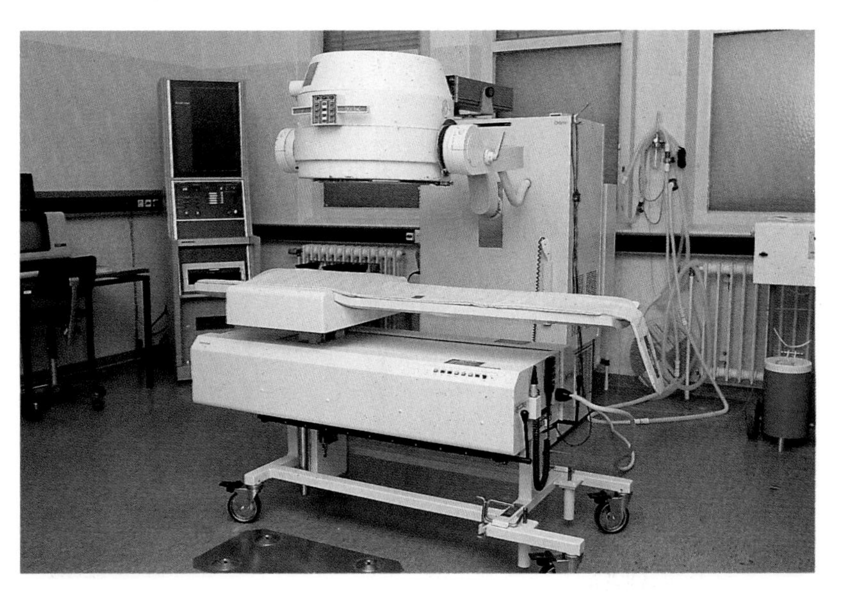

Figure 7.6a
Radioisotope scanning

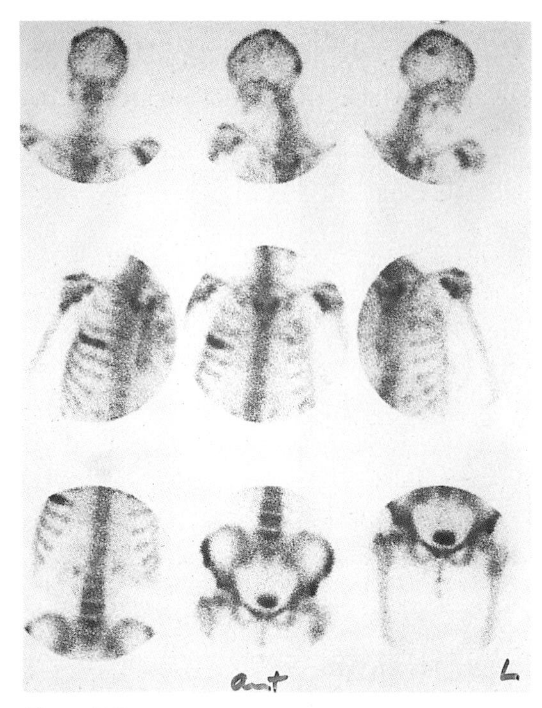

Figure 7.6b
Radioisotope scan

factor involved in the subsequent development of hepatic cancer.

4. Diagnosis of neoplasia

a. Clinical diagnosis

Clinical diagnosis by means of history taking and physical examination by observation, palpation, percussion and auscultation still provide a sound basis for diagnosis. There are also many advanced investigative techniques and only a few can be described briefly here.

b. X-ray examination

Traditional X-ray methods and the newer imaging techniques such as CAT (computer aided tomography) and MRI (magnetic resonance imaging) can help to localise tumours (Figure 7.5a and 7.5b).

c. Scanning after radio-isotope administration

Radioactive agents can be selected to accumulate in specific tissues or organs. A two dimensional scan by a radiation detector can thus be built up after a tiny, safe dose has been given to the patient (Figures 7.6a and 7.6b).

d. Endoscopy

Flexible fibre-optic scopes can now penetrate more and deeper recesses of the body

Common endoscopy examinations include:
- laryngoscopy;
- bronchoscopy;
- gastroscopy;
- cystoscopy;
- colonoscopy.

Figure 7.7
Oesophagoscopy

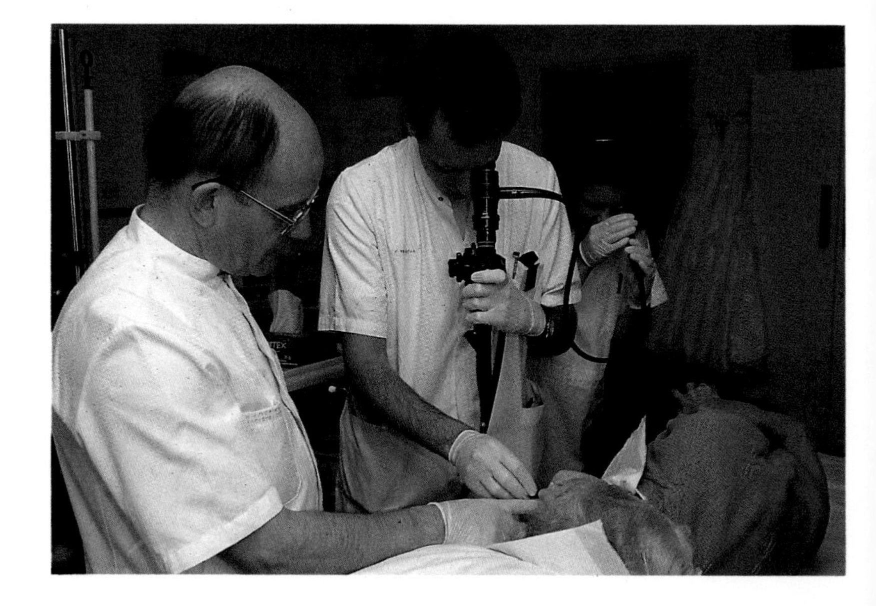

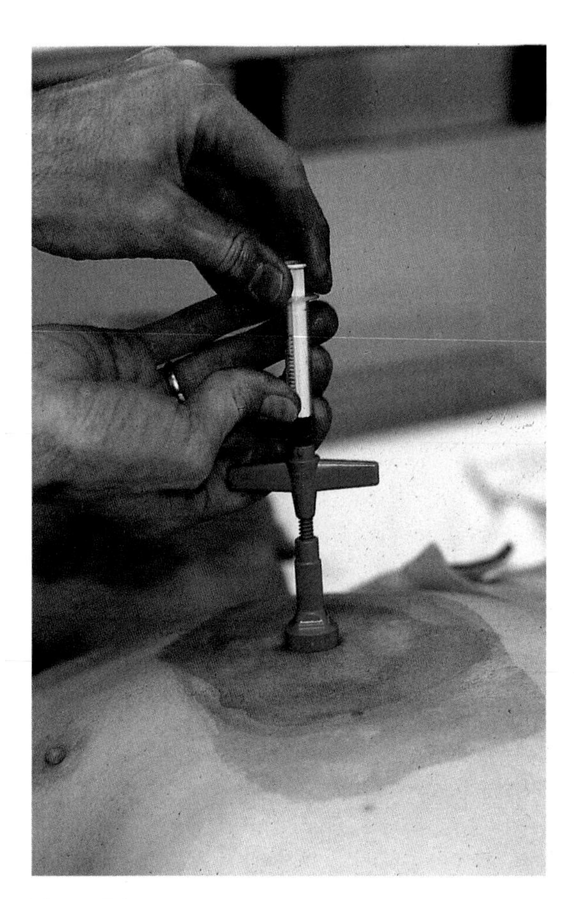

Figure 7.8
Sternum puncture

e. Biochemical examination

Some neoplasms produce specific tumour marker chemicals which can aid in diagnosis and determining therapy and prognosis.

f. Histological examination

A *biopsy* (i.e. a sample of body tissue) usually provides the most specific tissue diagnosis, by allowing microscopic or histological study of the structure of body cells and tissues.

g. Cytology

Examination of aspirated cells now provides more and more diagnostic information without the need to formally take a large excision biopsy. Haematological malignancies can frequently be diagnosed by sternal marrow aspirate examination. Examining cells obtained by such methods, rather than a whole piece of body tissue as in histology, is known as cytology. Early pre-invasive treatable changes can be recognised by cytological techniques such as cervical cytology.

5. Treatment

Details of cancer therapy are beyond the scope of this book, but we can briefly list some of the basic techniques.

Surgery
Curative or palliative.

Radiotherapy
Curative or palliative.

Chemotherapy

Immunotherapy
This is as yet experimental but may prove effective and tumour specific.

Hormonal therapy
Some tumours, such as prostatic carcinoma, show a response to sex hormones (oestrogens in the case of prostatic lesions).

Supportive therapy
Care of the psyche as well as the soma is necessary in what may be a long and, in some cases, terminal illness.

6. Summary

This chapter has provided a basic description of:
- the concept of neoplasia;
- some diagnostic techniques to detect neoplasia;
- an outline of the principles of treatment.

8 Clinical history taking and physical examination

1. Introduction

Although expensive high-tech investigations can give a great deal of useful information, a patient's own account of the symptoms still provides the most useful basis for a correct clinical diagnosis. After history taking, the physical examination of the patient is of vital importance in providing valuable clinical information. The results of the physical examination often determine whether further scrutiny is required (perhaps by a specialist) or what course of treatment should be followed.

Learning outcomes

After studying this chapter, the student should have sufficient knowledge and understanding of:
- the importance of a systematic clinical history;
- the problems and practice of history taking;
- the methods and techniques of physical examination;
- the basic information which may be obtained from physical examination.

2. History taking

A clinical history is primarily obtained from the patient. At times, for example if the patient is a young child, or is confused or comatose, a relative or friend will provide the necessary details. Successful history taking can only be completed if the interviewer (the nurse or the doctor) gains the confidence of an often worried and fearful patient.

Details of the patient's history must be systematically organised. General data such as full name, address, date of birth, gender, and occupation are important, not only to provide a check of identification for avoidance of a possible future mistake, but also to place the patient in the relevant perspective, some diseases being common in familial groupings and others being linked to specific occupations. The patient's current major complaint is then detailed, followed by a note of any secondary complaints which may or may not be related to the main problem.
A detailed history relating to all the bodily systems then follows.

An accurate history of previous illnesses and operations is recorded together with notes of any current drug and other therapy.

Past and present occupational details are

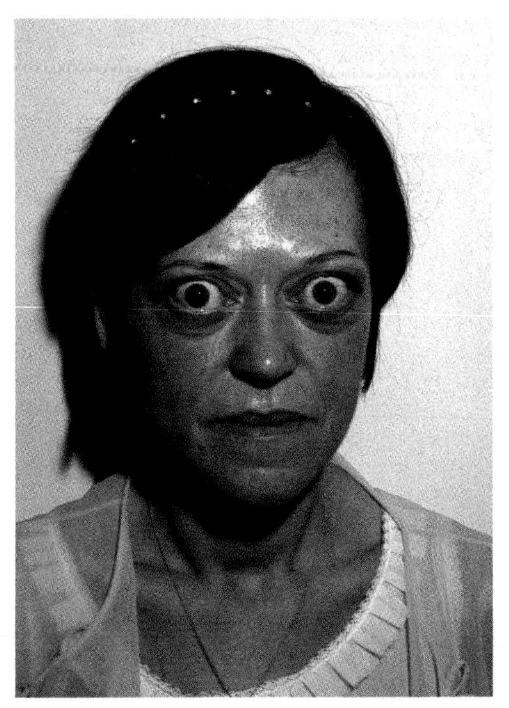

Figure 8.1
Exophthalmus

recorded, as is social information such as alcohol and cigarette consumption.
Details of how to present a specific systemic history are available in standard medical nursing textbooks.

3. Physical examination; methods and the data obtained

The elements of physical examination are:
- inspection;
- palpation;
- percussion;
- auscultation.

a. Inspection
Often underestimated, inspection is of prime importance.
A first meeting with the patient provides a general picture of posture, mood, nutrition, level of consciousness, and psychological demeanour.

Inspection of the skin and mucosal surfaces reveals alterations in colour such as:
- cyanosis;
- jaundice;
- anaemia;
- dehydration or oedema.

Breathing patterns can be observed. A good nose is valuable in detecting acetone in diabetics or alcohol, for example, in the patient's breath. The mouth gives clues to hydration, nutrition, and possible previous medication. The teeth (or their absence) give information regarding possible nutritional problems in the elderly. The eyes show any presence of jaundice.The conjunctiva give an indication of possible anaemia. The pupils can reveal various neurological defects.

Protrusion of the eyes (exophthalmos) in hyperthyroidism can easily be observed (Figure 8.1). Examination of the neck reveals thyroid swellings and lymph node enlargement (Figure 8.2).

Figure 8.2
Swollen lymph node in the neck

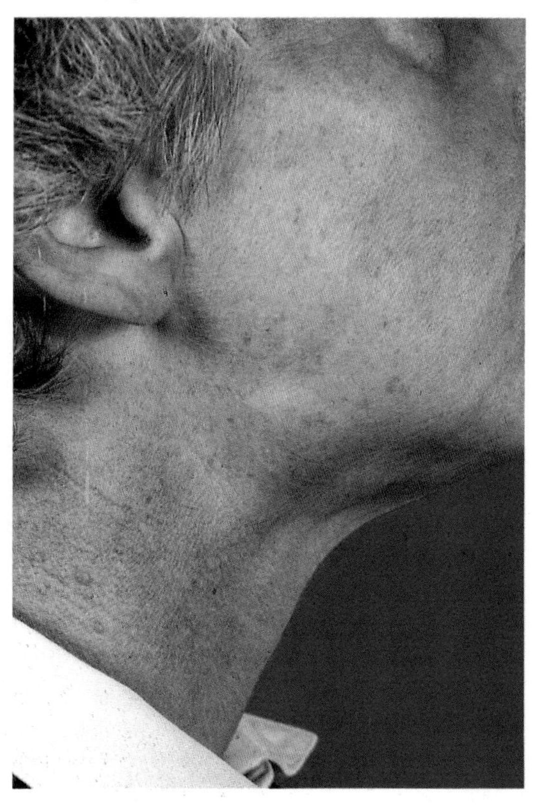

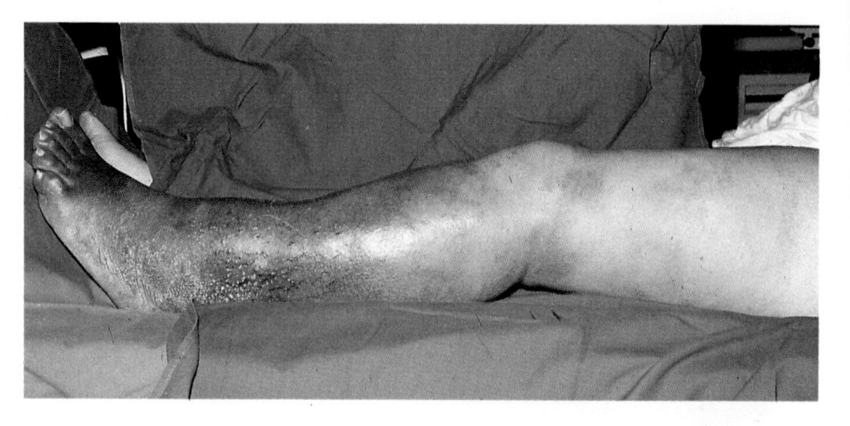

Figure 8.3
Leg with thrombosis and foot necrosis

Inspection of the thorax shows breathing patterns, evidence of the use of neck or abdominal muscle in respiration and any asymmetry.

Inspection of the female breast may reveal a hitherto unnoticed swelling. All previous surgical or other scars must be recorded.

Inspection of the abdomen shows reduced movement if the patient has abdominal pain. In peritonitis the abdomen does not move with respiration. Any swelling in the abdomen can be observed, and physiological causes such as fat, fluid, flatus, faeces or fetus, must be considered. An abnormal pulsation may indicate the presence of an aortic aneurysm. Examination of possible hernial sites should be recorded. Large testicular tumours have been tragically missed due to prudery on the part of doctor or patient. The limbs should be observed at rest and in motion.

Inspection of the legs will reveal details of the circulatory system, including oedema, peripheral ischaemia, gangrene, leg ulcers or venous obstruction (Figure 8.3). 'Bed-sores', if present, need careful attention. Leg ulcers should be recorded (see Figure 3.4). Chronic respiratory disease may be manifested by the presence of finger or toe 'clubbing' (Figure 8.4a and b). Arthritis can often be detected by inspection of fingers and toes. The nails provide a valuable amount of data about the patient's general health, and can show conditions such as fungal infections, anaemia, circulatory disorders, nutrition and occupational trauma. Toe deformities such as hallux valgus are easily noted (Figure 8.5).

Inspection reveals any limb deformity whether congenital, chronic or acquired, such as a recent fracture or dislocation.

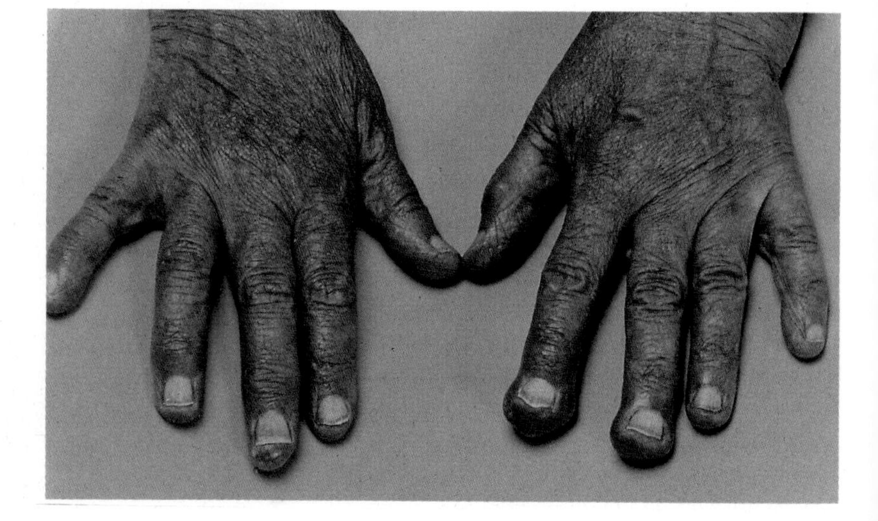

Figure 8.4a
Club fingers (back of hands)

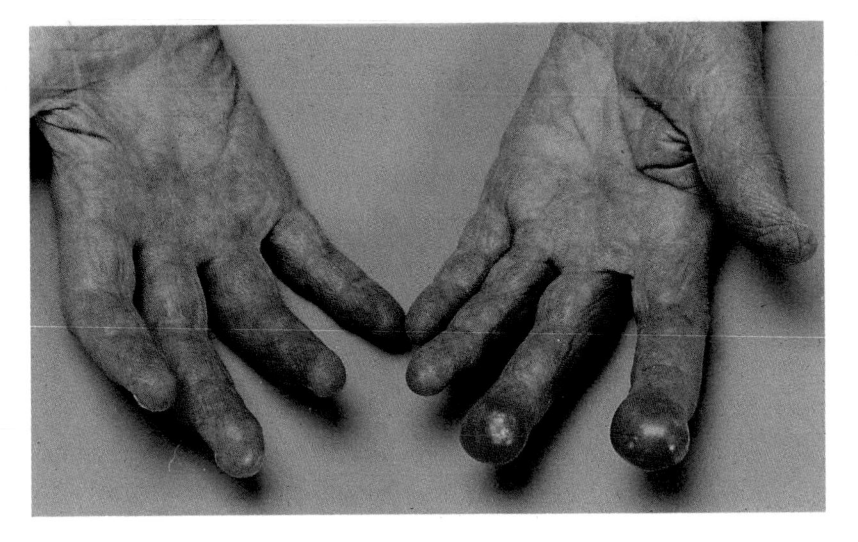

Figure 8.4b
Club fingers (palm of hands)

b. Palpation

Examination by palpation, although primarily the province of the physician, is important for all health care professionals. Any lump or bump should be recorded (Figure 8.6). An observant nurse may be the first to palpate an enlarged lymph node or haematoma.

The pulse can be palpated in several anatomical sites, and details of its examination are considered in a separate chapter. Internal digital examination of the rectum and vagina may provide the examining physician with evidence of possible tumour and other lesions.

Figure 8.5
Hallux valgus

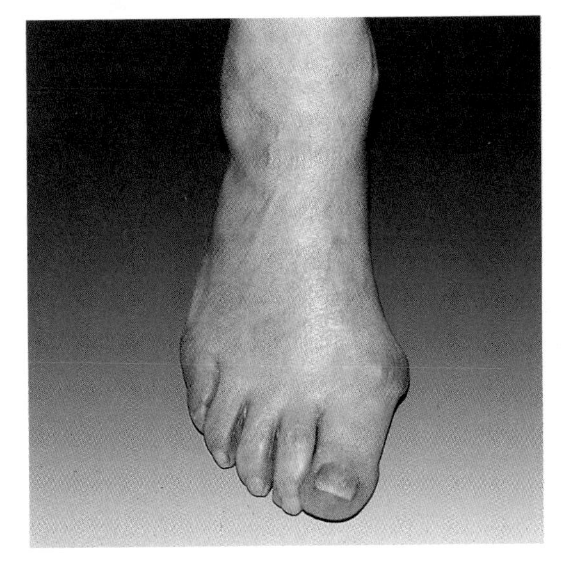

Figure 8.6
Palpation of the liver

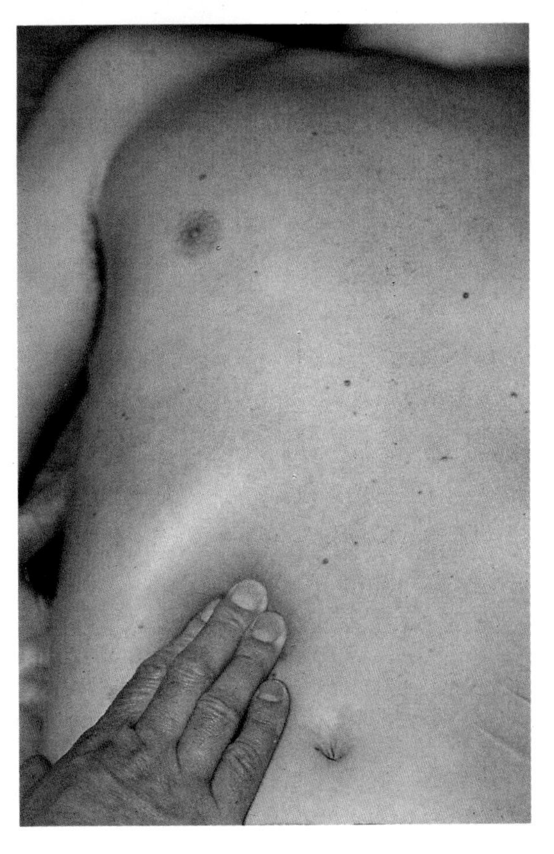

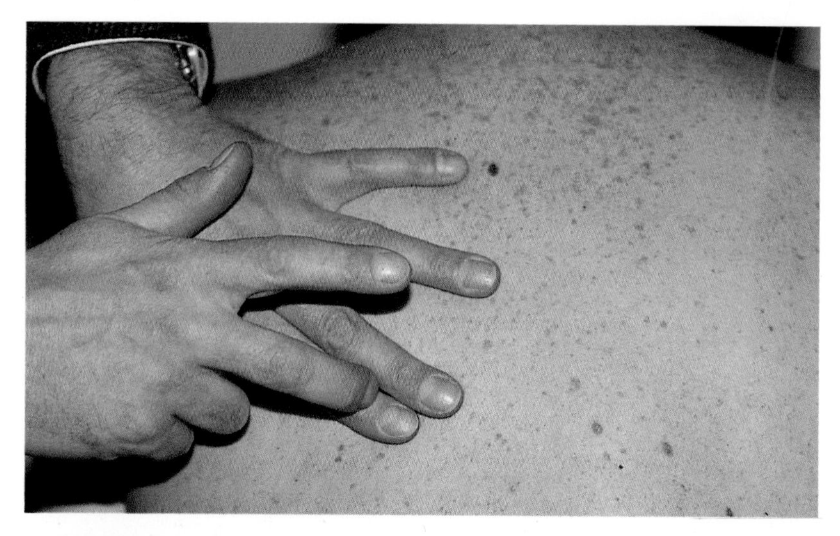

Figure 8.7
Percussion

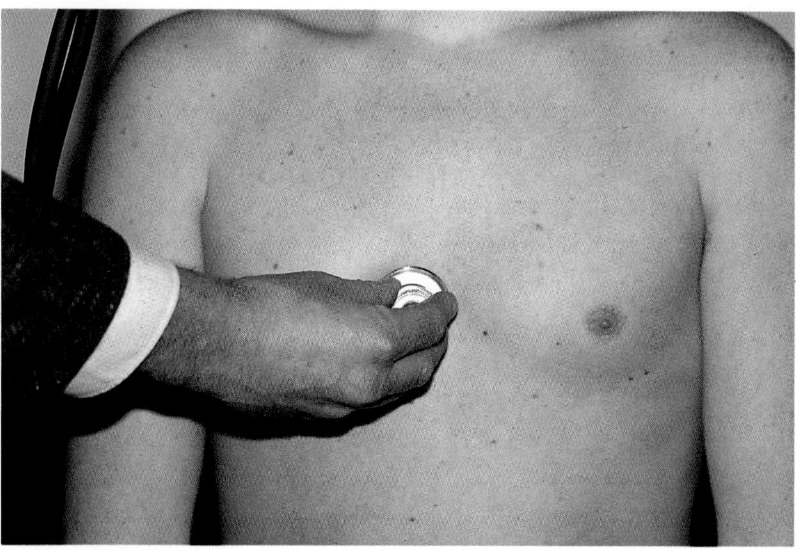

Figure 8.8
Auscultation

c. Percussion

Percussion determines the size and position of the heart, lung consolidation or pleural effusion or pneumothorax (Figure 8.7). An enlarged bladder can be percussed, as can a distended intestine.

d. Auscultation

The doctor gains diagnostic information from auscultation of the heart, lungs, certain vessels and the intestines. A stethoscope, a most elementary device, is still an indispensable clinical tool (Figure 8.8).
Careful auscultation is also essential for the accurate measurement of blood pressure.

4. Summary

A full and detailed history of the patient will provide a valuable basis for forming an accurate clinical diagnosis. Physical examination of the patient consists of inspection, palpation, percussion, and auscultation.

Body temperature

9

1. Introduction

Recording a patient's temperature is a routine activity for every nurse. The result, however, is vital in the management of many illnesses. Changes in temperature may represent either the occurrence of a complication or an improvement in the disease process.

Learning outcomes

After studying this chapter, the student should be able to:
- explain temperature homeostasis;
- describe varieties of abnormal temperature patterns;
- define hypothermia.

2. Body temperature range

Normal human body temperature is 37 degrees Celsius. At this temperature, normal metabolic biochemical activity is efficient.
The surrounding temperature of the environment is usually lower than 37 degrees, so the body needs an efficient mechanism for internal heat control. Body temperature is mainly controlled by the skin. By variation in blood supply to the skin, heat exchange is controlled. *Sweating* leads to rapid heat loss by evaporation when necessary. *Shivering* reduces skin blood supply and so reduces heat loss. Temperature receptors, specific neural detectors in the skin, can rapidly adjust the cutaneous blood supply and hence govern the body's temperature. If this mechanism proves insufficient, the body temperature may rise, resulting in fever (pyrexia), or fall, resulting in hypothermia. Some central nervous system control persists, however, and to keep the body alive its 'thermostat' will be temporarily reset at a different level. In cases of haemorrhage into the part of the brain called the pons, central control is lost and the patient's temperature rises inexorably. This condition is referred to as malignant hyperpyrexia, and death rapidly intervenes.

An elevated temperature may cause cerebral irritation. This is called delirium.

3. Measurement of body temperature

The traditional glass clinical thermometer is gradually being replaced by electronic temperature transducers. Clinically, a rectal temperature gives the most accurate record of core temperature, but local disorders such as haemorrhoids, rectal fissures, or obesity, may make this method difficult, and it is also difficult with children. In forensic post-mortem cases, core temperature is often measured on

Figure 9.1
Some types of abnormal temperature course:
1. Subfebrile
2. Continuous fever
3a. Intermittent fever
3b. Remittent fever

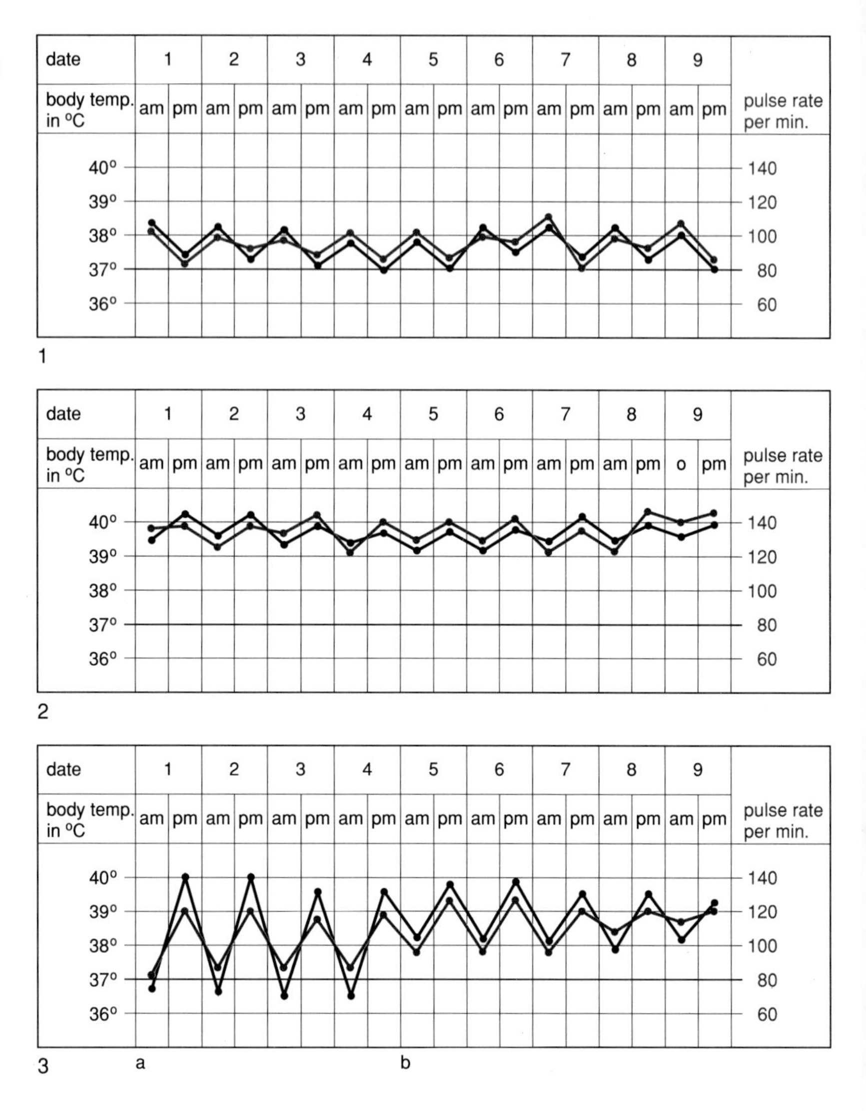

several occasions to plot a decay curve and thus establish a time of death.

Axillary temperature is usually 0.5 degrees Celsius lower than oral or rectal temperature.

4. Types of temperature variation

An elevated whole-body temperature is referred to as fever or pyrexia. An elevation in temperature to not more than 38 degrees Celsius is referred to as a *sub-febrile temperature* (Figure 9.1.1). A prolonged elevated temperature at or over 39 degrees Celsius showing little variation is called a *continuous fever* (Figure 9.1.2). Such a continuous fever indicates serious systemic infection or damage to pontine temperature control centres.

An *intermittent fever* (Figure 9.1.3a) is one in which an elevated temperature falls to normal or near normal and rises again within a 24 hour period.

A *remittent fever* (Figure 9.1.3b) shows a variation of 1 or 2 degrees but does not reach a near normal temperature within a 24 hour period.
Temperature patterns can be specifically diagnostic in deciding types of malaria.

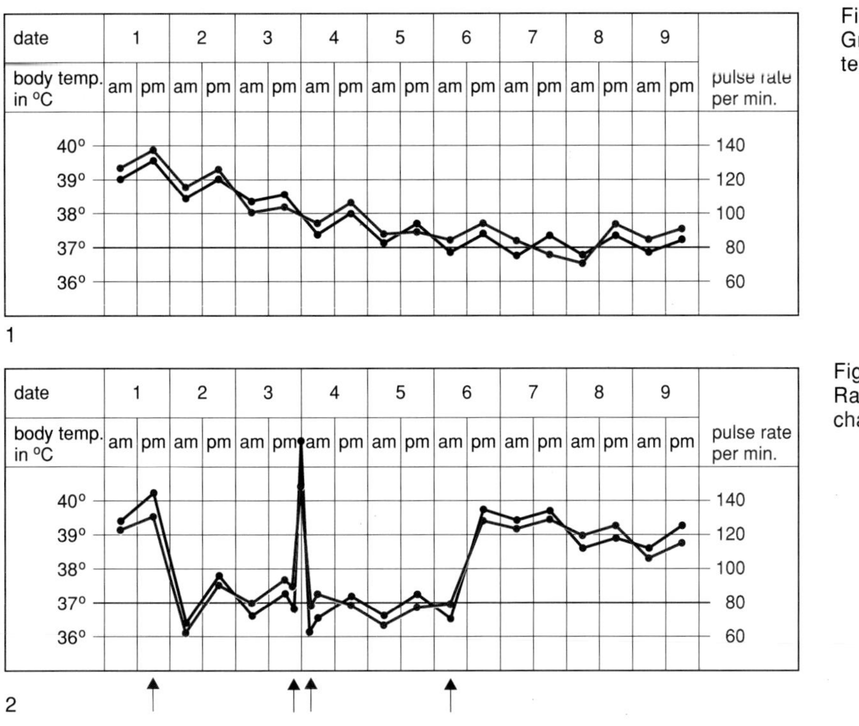

Figure 9.2.1 Gradual fall in temperature

Figure 9.2.2 Rapid temperature changes

5. Signs and symptoms observed in fever

A gradual modest rise or fall in temperature causes few perceived symptoms apart from fatigue and apathy (Figure 9.2.1).
Sudden rises in temperature (Figure 9.2.2) usually result in shivering, teeth chattering and a dry, pale skin, as hypothalamic central nervous controls attempt to reset the body's 'thermostat' and achieve homeostasis as quickly as possible.

When the opposite occurs, and the temperature suddenly drops, the patient perspires freely and the skin becomes hot and red.

6. Body temperature and pulse rate

There is a well recognised relationship between temperature and pulse rate. A rise in temperature of one degree results in an increase in pulse rate of ten to twelve beats per minute (for more information on pulse rate, see Chapter 10).

An elevated pulse rate without an elevated temperature is, of course, seen in many conditions (Figure 9.3). Increased pulse rate is known as tachycardia and this is seen physiologically after exercise and in many pathological conditions, including heart disease, thyroid over-activity and states of heightened emotion.

Relative bradycardia (slow pulse rate) is observed when the body temperature is higher than would be expected from the pulse rate (Figure 9.4). This is seen in some viral infections, certain diseases of the central nervous system, and at times in cases of jaundice.

7. Subnormal body temperature (hypothermia)

A patient can be recognised as being in a state of hypothermia when the core temperature drops below 35 degrees Celsius.

Hypothermia kills otherwise fit young adults every winter in Europe when ill-equipped expeditions are made in mountainous or

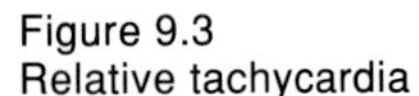
Figure 9.3
Relative tachycardia

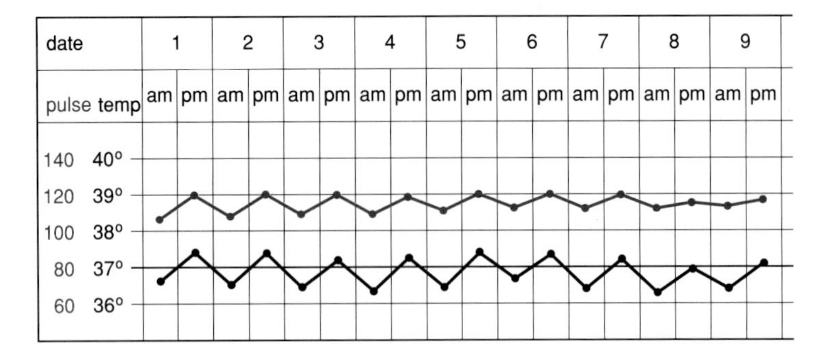

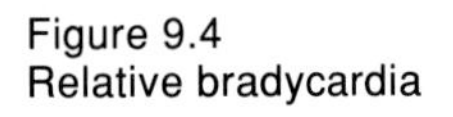
Figure 9.4
Relative bradycardia

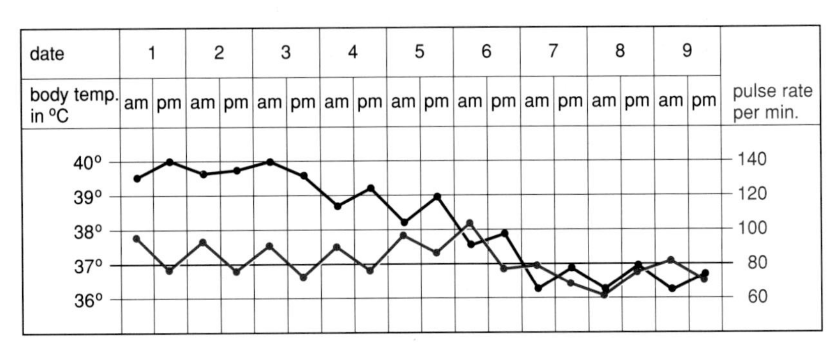

similarly hostile locations. The elderly are prone to hypothermia especially as a secondary phenomenon after unattended cerebro-vascular accident, diabetic coma, or alcoholic intoxication.

In young infants and in the elderly a serious infection may paradoxically lead to a temperature drop.

8. Summary

This chapter has described how the body controls temperature. Temperature measurement has been discussed briefly. Specific patterns of fever have been outlined and hypothermia has been discussed.

The pulse

10

1. Introduction

The pulse shows six characteristics which individually and together provide a host of important clinical information not only about the cardiovascular system but also relating to many other clinical states. These characteristics are:
- pulse rate or frequency;
- pulse regularity or rhythm;
- equality of pulse force;
- pulse volume;
- pulse force;
- condition of the vessel wall.

Learning outcomes

After studying this chapter, the student should have sufficient knowledge and understanding of:
- pulse, tachycardia, bradycardia, fibrillation and arrhythmia;
- the connection between pulse wave and heart action;
- the changes in heart rate;
- the connection between regularity and equality;
- the consequences of disease processes affecting the vessel wall.

2. Rate or frequency

The frequency of the left ventricular contraction determines the pulse rate. In normal circumstances this rate lies between 60 to 80 beats per minute in healthy young adults (Figure 10.1a). It is higher in children because their energy needs are higher. Physical exercise or emotion will also raise the pulse rate in otherwise healthy adults. A rapid pulse rate of more than 100 beats per minute is called a *tachycardia* (Figure 10.1b).

Tachycardia is seen in many disease processes, including over-activity of the thyroid gland, states of fever, and any systemic disease involving tissue necrosis, inflammation, or malignancy. In these conditions the cardiovascular systems are compensating to adjust for the need for more oxygen.

A more or less comparable situation is seen in cases of anaemia where there is insufficient haemoglobin to bind adequate oxygen for the body's needs. This can be to some extent compensated by an increased pulse rate leading to faster blood circulation.

Tachycardia is seen in heart disease where the pumping force of the myocardium is diminished, for example in heart failure due to myocardial damage or cardiac valve disorders. In these cases the heart is attempting to pump an effectively smaller volume of blood around the circulatory system faster, to allow sufficient oxygen to reach the tissues.

In cases of shock too, where there is an insufficient circulating volume, the heart rate will increase in an attempt to maintain minimum oxygen supplies to vital organs. Reflex compensatory tachycardia is therefore a useful homeostatic mechanism enabling vital organs such as the brain to maintain a sufficient oxygen supply for survival.

Where the normal heart rhythm is disturbed, the resultant tachycardia is not a homeostatic response and is not a useful mechanism of adaptation. In such cases the heart rate often exceeds 120 beats per minute and at times may even exceed 200 beats per minute. If the rhythm is totally irregular and rapid a state of *fibrillation* exists. If the source of this fibrillation is atrial, then the condition is referred to as atrial fibrillation. If the source is ventricular, the resultant ventricular fibrillation will rapidly prove fatal unless corrected. This can be achieved by emergency electric shock conversion to a more stable rhythm pattern (*defibrillation*).

Chapter 9 has already described relative tachycardia, where the pulse rate is higher than might be expected from the body temperature. *Bradycardia* is a slow pulse rate of less than 50 beats per minute (Figure 10.1c). Sometimes the occurrence of a slow pulse rate is physiological, as in sleep or in a trained athlete.

In sub-thyroid diseases the pulse rate and the temperature are lowered as a reflection of reduced metabolic rates in the body. A very low pulse rate is seen in cases of hypothermia, sometimes so low that sufferers have inadvertently been declared dead. Vagus stimulation also results in a slow pulse rate, and this is seen in disorders of the central nervous system including tumour, haemorrhage, thrombosis, or inflammation, where it is caused by pressure on the cerebral tissues. A lowered pulse is sometimes seen in typhoid fever, some viral infections, and in obstructive jaundice. Bradycardia is seen in patients on digoxin therapy and careful monitoring of the rate is essential to detect early overdose, since the therapeutic dose is close to a toxic dose. A slow pulse rate in relation to that expected from the body temperature is referred to as *relative bradycardia*.

A very slow heartbeat of less than 40 beats per minute is seen in cases of heart block (Figure 10.1e). The conduction of stimuli from the sino-atrial node to the ventricle is blocked by a lesion affecting the bundle of His (or atrioventricular bundle). Sometimes the impulse in the sino-atrial node is itself blocked. This is seen fairly frequently in the elderly.

3. Regularity

In normal circumstances the pulse is more or less regular, which means that the intervals between successive beats are equal.

The regularity of the pulse rate can be checked by beating time to each beat. A very precise assessment of this regularity, however, presents a real problem. The rhythm observed by monitoring of the pulse is a reflection of the ventricular contractions, which do not always follow a steady rhythm. They are dependent on the volume of blood which is pumped during each beat, and the volume of blood available at the end of inspiration exceeds that available at the end of expiration. The pulse rate is thus a little faster at the end of inspiration than at the end of expiration, but the difference is so small as to be only detectable by pulse palpation if the breathing is extremely deep. This normal physiological variation is referred to as respiratory arrhythmia.

The commonest and most-easily recognised totally irregular pulse is found in *atrial fibrillation* (Figure 10.1d). When this occurs, multiple impulses are generated in the atrium at such a rate that the ventricles can only respond irregularly to a few of them. These impulses, conducted in a random fashion, cause a rapid and irregular contraction of the ventricles and the pulse is therefore fast and irregular.

An entirely different phenomenon which can lead to an irregular pulse is an *extrasystole*. This is a premature ventricular contraction caused by

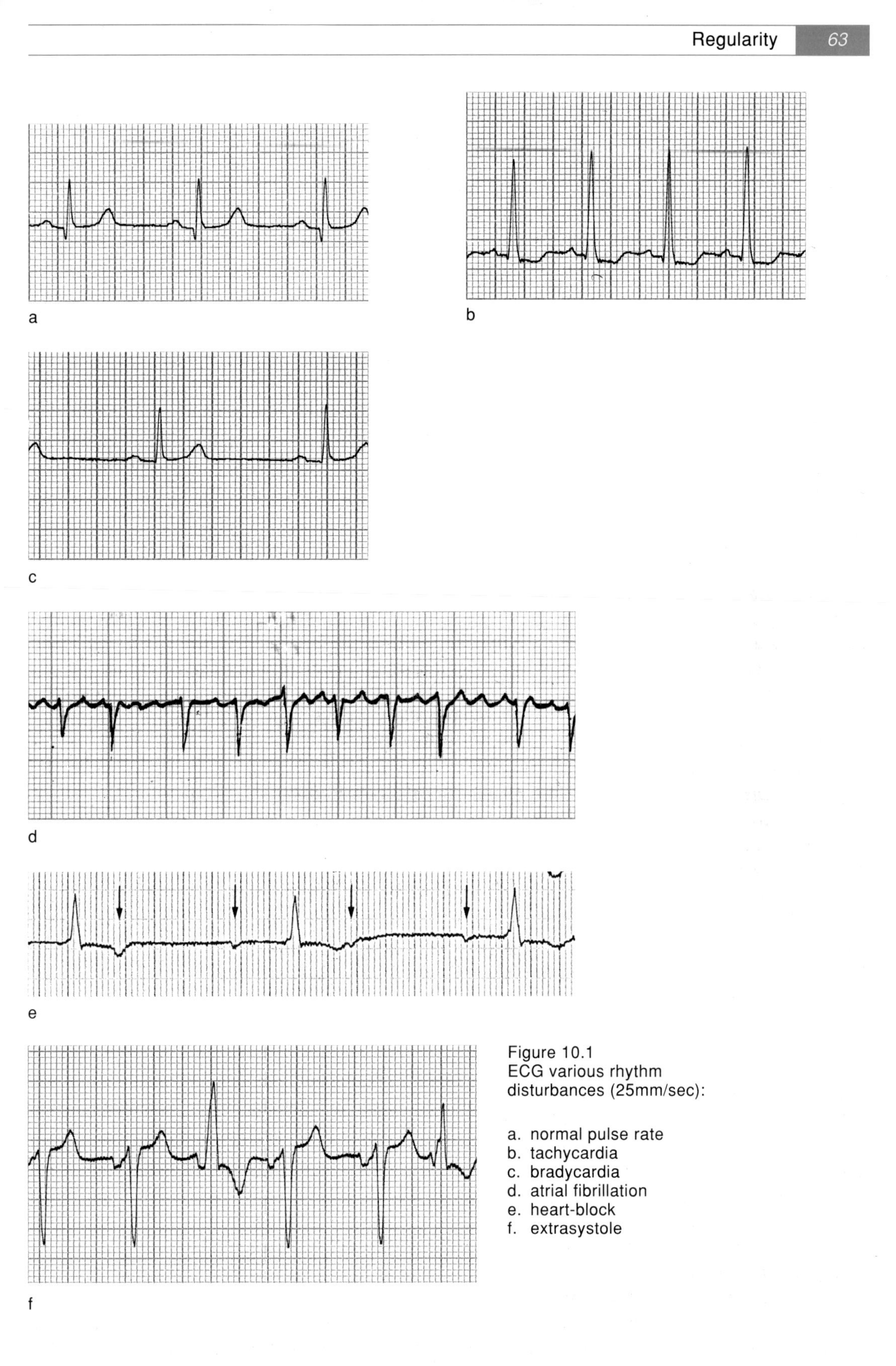

Figure 10.1
ECG various rhythm disturbances (25mm/sec):

a. normal pulse rate
b. tachycardia
c. bradycardia
d. atrial fibrillation
e. heart-block
f. extrasystole

a premature stimulus to atrium or ventricle. The early ventricular contraction occurs when the left ventricle is incompletely filled with blood, leading to a perceived weak beat. The subsequent beat appears forceful as the ventricle attempts to compensate (Figure 10.1f).

There are two types of extrasystole; the ventricular extrasystole and the supraventricular or atrial extrasystole (Diagram 10.1). In a ventricular extrasystole, the anticipated subsequent heartbeat is delayed by a compensatory apparent missed beat.

The extra long interval between the extrasystole and the subsequent heartbeat is called the compensatory interval. The experience of the patient having a ventricular extrasystole is that the heart has stopped and starts again with a jerk. In a supraventricular extrasystole the subsequent heartbeat is felt at an expected interval and the patient is therefore unaware of the extrasystole.

4. Equality of pulse force

Normally, in each ventricular contraction the volume and force of contraction are virtually the same as in the previous one and the subsequent one. The volume of blood moved through the arteries per beat is more or less identical. If the heart muscle (myocardium) is diseased, the heart will have difficulty in pumping equal volumes of blood at each contraction and beats of different force will be observed. Inequality of the pulse force may thus be expected as a result of irregularity of the pulse. The volume of blood pumped at each ventricular contraction also varies, so that the pulse which is irregular in rhythm is also unequal in force, whether the cause is atrial fibrillation or extrasystole.

5. Volume

Pulse volume reflects the amount of blood pumped by the heart per beat or, more precisely, the volume of blood moving with each pulse wave through the artery under palpation. As previously noted, in atrial fibrillation the volume of blood varies considerably. It should also be noted, however, that in states of shock when there is too little available blood in the vascular system the heart beats rapidly,but the volume per beat is small.

6. Force

Palpation assesses the force of systolic contraction of the left ventricle. A healthy heart is recognised by a pulse which feels strong.

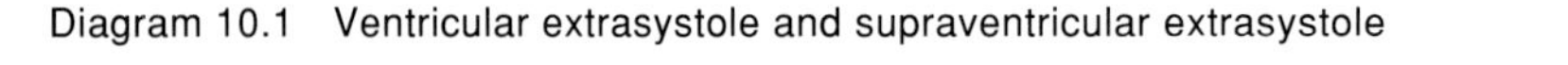

Diagram 10.1 Ventricular extrasystole and supraventricular extrasystole

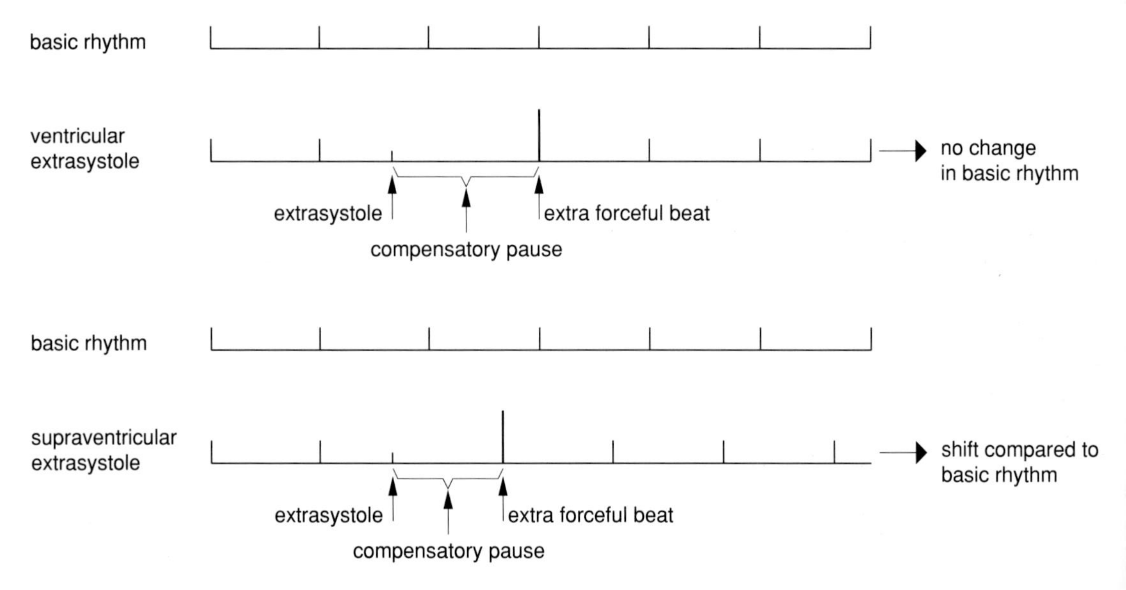

Clinical appreciation of the force of an arterial pulse is very unreliable. Observation of blood pressure by a sphygmomanometer provides a more accurate and useful measurement.

7. Condition of the vessel wall

The elasticity of the arterial wall decreases with the normal ageing processes. The vessel wall can become more fibrous, causing arteriosclerosis. Degenerative changes such as atheroma cause depositions of fatty materials followed by fibrosis and sometimes by calcification. This condition is also known as atherosclerosis. The artery as palpated seems crooked and rigid. Such arteries may occlude readily with a thrombus (blood clot), as in coronary artery thrombosis, leading to distal ischaemia (tissue damage due to lack of blood supply) and infarction (necrosis or tissue death due to lack of blood supply). In the legs, femoral artery occlusion leads to distal ischaemia of the foot and leg and, if total, to gangrene. Arteries severely affected by occlusive atheroma include the coronary arteries, femoral arteries, carotid arteries, cerebral arteries, and sometimes the mesenteric arteries supplying the intestines.

Atheroma of the abdominal aorta may lead to mechanical weakening and an expanding aneurysm will develop. This can rupture with massive and often fatal retroperitoneal haemorrhaging.

8. Summary

Palpation of the radial pulse still provides an excellent assessment of cardiac function. This chapter has described the properties of the pulse in terms of:

- rate;
- regularity;
- equality of force;
- volume;
- pressure;
- condition of the vessel wall.

11 Blood pressure

1. Introduction

In layman's language, the term *blood pressure* implies the pressure with which blood travels through the cardiovascular system, including the arteries, the veins, and the capillary bed. In medical parlance, however, the term blood pressure generally refers to the recorded pressure in major arteries.

Learning outcomes

After studying this chapter, the student should be able to:
- explain the origin of blood pressure;
- describe the way in which blood pressure is measured;
- identify the limits of normal blood pressure;
- explain varieties of hypertension and hypotension;
- outline some causes of raised venous pressure.

2. Blood pressure and cardiac function

Blood pressure is virtually identical to the pressure with which the left ventricle ejects blood into the aorta at every contraction (*systole*). This high pressure, contrary to initial impressions, diminishes only slowly after relaxation (*diastole*) of the left ventricle's myocardial contraction, and the residual pressure remains considerably above zero due to the fact that blood cannot travel quickly through the peripheral vascular bed; it encounters peripheral resistance because of the small calibre of the peripheral vascular channels. Being very delicate, these distal small vessels could not withstand the direct high pressure coming from the left ventricle and the aorta.

The aorta is anatomically designed to buffer the pressure of the ventricular output, and its wall, when healthy, is protected with an elastic structure. This elastic mechanism enables the aorta to expand considerably and in essence store a large volume of the left ventricular output. When the left ventricle relaxes in diastole, the high pressure in the aorta is reduced and the stretched wall can then gradually relax and transmit the blood volume to the periphery of the system in a more controlled manner. When it is time for the next ventricular contraction (*systole*), the aorta has regained its original shape and volume. The peripheral pressure in arteries is also prevented from decreasing to zero by the many small channels and branches in the peripheral circulatory system. These small arteries (arterioles) are largely responsible for the variable distribution of blood to the periphery through their ability to contract or relax selectively. In anaemia, when a large volume of blood is needed to maintain oxygen supply, the arterioles dilate. When the blood pressure in the arteries is higher than normal, increased contraction in muscular arterioles prevents the high pressure blood from reaching and damaging small and vulnerable peripheral capillaries.

The lowest level achieved by arterial pressure is determined by the balance between the elasticity of the aorta and the resistance of the arterioles as determined by their degree of contraction. This pressure is referred to as the *diastolic pressure. Systolic pressure* is the highest blood pressure level immediately after left ventricular contraction. The difference between these two pressures determines the volume of blood travelling to the tissues. The peripheral supply is therefore directly dependent on this difference between the peak systolic pressure and the lowest diastolic pressure, which is called the *pulse pressure,* and is the pressure felt when the pulse is palpated.

3. Arterial blood pressure

a. Measurement of arterial blood pressure

In normal clinical practice arterial blood pressure is measured with a *sphygmomanometer.*

The measurement is carried out by applying a cuff to either the right or left upper arm. Pressure is then applied slowly to the cuff by inflation. This cuff is attached to a tube containing mercury calibrated against a scale marked in mm of mercury. A stethoscope applied over the brachial artery is then utilised to listen to the pulse wave in the artery transmitted from the left ventricle.

Firstly the cuff is inflated to a pressure of approximately 200 mm of mercury. The pressure is then slowly and steadily reduced at a rate of about 2–5mm per second. At the precise moment when the *systolic blood pressure* is reached, the peak of the pressure wave above the cuff will be able to reach to below the cuff for a short interval during each ventricular contraction. This can be heard through the stethoscope as a kind of short tapping sound. As the air in the cuff is released the pressure continues to fall until a pressure is reached when the cuff is no longer an obstruction to the blood flow. When there is steady flow underneath the cuff the transmitted vascular sound wave disappears. This level is the *diastolic blood pressure.* By tradition the observed diastolic blood pressure is recorded at the point where the tapping sound first obviously changes from a clear to a muffled sound.

Standardisation of technique by applying the cuff at heart level on the arm is recommended. In very obese patients a wider than usual cuff should be used otherwise an erroneously high pressure will be recorded (up to 25 mm mercury too high).

The cuff can be secured by simply tucking in the end, by hooks or by velcro. A more accurate measurement is recorded if an anatomically tapered cuff is used. Initial measurement may be unreliable (usually too high) due to patient anxiety, and repetition should help. A patient undergoing repeated long term blood pressure monitoring should have the procedure carried out at the same period each day, and he should also be relaxed if the results are to be meaningful.

The sphygmomanometer must be regularly calibrated and the cuff checked and replaced if worn. If the device is broken great care must be observed, as mercury is very toxic.

b. Normal blood pressure

In the aorta and in the major arteries the blood pressure is normally at a constant value of 120–130mm of mercury.

This means that the heart is effectively able, 70 times a minute, to raise a column of mercury 120–130mm. This equates to a column of water 1.75m in height more than once a second – it is a remarkable achievement for any pump; it is especially so for a pump which is only the size of a fist, and which must perform continuously and without time out for service or repair for a period of seventy years or so.

The normal blood pressure increases slightly with age, as can be seen from Table 11.1. An acceptable blood pressure for an elderly person may be too high for a young adult, and in general terms a normal systolic pressure is acceptable if measured as 100 mm mercury plus age in years, but only up to a limit of about 150 mm mercury. The diastolic blood pressure also rises slightly with age, from about 70 mm mercury in young adults to an acceptabele level of about 90 mm mercury in elderly people. The World Health Organisation has defined the normal limits as 150 mm mercury for systolic blood pressure and 90 mm mercury for diastolic blood pressure. Higher values are referred to as *hypertension*.

Age	Systolic	Diastolic
15 – 20 years	90 – 120	60 – 80
20 – 30 years	100 – 130	70 – 90
30 – 50 years	110 – 140	70 – 90
> 50 years	120 – 150	70 – 90

Table 11.1
Summary of normal blood pressure levels.

It must always be remembered that the first measurement taken may be 15–20mm mercury too high for the systolic pressure, and 10–15mm mercury too high for the diastolic.

c. Fluctuations in arterial blood pressure

Hypertension

Regularly achieved levels of blood pressure in excess of those defined above as normal constitute a state of *hypertension*. For the diagnosis of hypertension the diastolic blood pressure is at least as important as the systolic, and probably more so. The blood pressure depends on the force of the heartbeat, the peripheral resistance, the vascular volume, and the elasticity of the aorta. The peripheral resistance, which reflects the degree of contraction of peripheral muscular arterioles, determines in the main the diastolic pressure.

The elasticity of the aorta and the force of the heartbeat largely determine the systolic pressure. Cases of hypertension in the absence of renal or other systemic disorders are referred to as cases of *essential hypertension*.

Blood pressure may be temporarily raised as a physiological response to exercise or emotion and it will return to normal levels when homeostasis is achieved. Sometimes there is an abrupt and abnormal rise in blood pressure due to the release of vasopressor hormones into the blood. This is called *paroxysmal hypertension* and it can be seen, for example, in cases of adrenal tumours.

The gradual increase in systolic blood pressure with age is caused by the loss of elasticity of the aorta as a result of atherosclerosis. But this does not constitute true hypertension, as the diastolic pressure remains unaltered.

This is also true for the elevation in pulse pressure caused by simultaneous increase in systolic, and decrease in diastolic pressure in aortic valve insufficiency, anaemia and hyperthyroidism. In severe myocardial disorders of a sudden onset (as after myocardial infarction), the systolic pressure may fall markedly despite pre-existing hypertension. In such cases the diastolic pressure may remain almost unchanged. This is referred to as *decapitated hypertension*, and such a condition is invariably serious.

Hypotension

The existence of physiological 'hypotension' remains debatable. A normal variation is of course observed in any population. Under the influence of emotions such as anger, fright or fear, peripheral vessels may suddenly dilate, causing the blood pressure to fall suddenly and markedly. This probably results from vagal stimulus. The patient appears pale, is dizzy and may lose consciousness transiently (in other words, he or she faints).

Recovery can be quickly brought about by lying the individual down and elevating the legs to aid venous return. Sudden standing after a period of lying or sitting may cause sudden transient drop in blood pressure. This is referred to as *orthostatic* or *postural hypotension*, and it is not unusual in the elderly and in pregnant women. In normal situations the autonomic nervous system can compensate for a change in posture by a corresponding change in peripheral resistance in the arterioles. If this mechanism is overwhelmed, approximately 750 ml of blood will temporarily be unavailable to the heart for output, bringing about a transient drop in blood pressure. The blood is, inappropriately, present in the arterioles.

A drop of more than 20 mm mercury systolic and 10 mm mercury diastolic causes symptoms of dizziness and vascular disturbance, such as black spots in front of the eyes. Some patients may even fall or lose consciousness.

Orthostatic hypotension may also be found in some diabetics and as a side effect of some medications.

Hypotension is also a feature of shock. In shock there is not only a sudden marked fall in blood pressure but also a marked increase in pulse rate. The patient's skin appears pale and clammy, the pulse pressure (the difference between systolic and diastolic pressure) is reduced, and the pulse as palpated feels weak and 'thready'.

After myocardial infarction the drop in blood pressure may be so severe as to be life threatening, and is indeed often fatal.

4. Venous blood pressure

Central venous pressure (CVP) is the pressure in the major veins. Generally, it fluctuates slightly but averages at around 0cm water. Near the heart, levels of minus 5 to minus 8 cm water are not abnormal. Initially, it might be anticipated that the venous pressure would be even lower as there is no obvious force providing pressure in the venous system. Three factors, however, act to create the central venous pressure. These are:
a. *The muscle pump.* At each muscle movement, contraction pushes blood in a venous direction.
b. *The heart* functions as a vacuum pump. On each ventricular contraction the atrial pressure falls and consequently the pressure falls in the supplying large venous channels.
c. *Respiration* also provides a suction effect. In inspiration the sub-diaphragmatic pressure rises relative to that in the chest above the diaphragm. This causes blood to flow from the abdomen towards the heart in the venous channels.

Venous pressure can be measured by an external or internal transducer monitoring the neck veins. A marked increase in central venous pressure is found in cases of congestive cardiac failure from whatever cause. External pressure on the superior vena cava, for example by neoplasm, leads to a more locally determined cause of increased central venous pressure.

5. Summary

Blood pressure is determined by the force of the heart beat, by peripheral resistance, by vascular volume, and by the elasticity of the major vessels. The difference between the systolic and diastolic blood pressure is called the pulse pressure.

Blood pressure measurement by sphygmo-manometer must be carried out under strict criteria in order for it to have meaningful results. Hypertension exists when the systolic pressure exceeds 150mm mercury and the diastolic pressure exceeds 90mm mercury. Essential hypertension exists when no causative factor can be determined.

Increased central venous pressure is observed in cases of cardiac decompensation.

12 Water balance

1. Introduction

The human body consists mainly of water in which numerous substances are dissolved. This water is found in the cells, in the intercellular fluid and in lymph. It is also found in blood which is, however, rich in cells (approximately 45–50%). All transport mechanisms and metabolic processes take place in an aqueous environment, and human life is impossible without water. Death always occurs from dehydration rather than from starvation. The control of the fluid balance is therefore a paramount bodily necessity. The *internal water balance* is the control of the fluid balance between the contents of the cells on the one hand, and the surrounding extracellular fluids on the other. The greater part (70%) of the *body fluid* is contained within the cells and is rich in sodium. Extracellular fluid (30%) is found outside the cells, and is rich in potassium. This difference in ionic composition is important for the function of many transport mechanisms. A shift in the water balance therefore involves an important shift in these salts or electrolytes, and a more accurate description of the scope of this chapter might be *water/electrolyte balance*.

Learning outcomes

After studying this chapter, the student should be able to:
- explain the concept of internal water balance;
- explain the concept of external water balance;
- describe the mechanisms which regulate water balance.

2. Regulation

Continuous and accurate regulation of fluid balance (Figure 12.1) is indispensable in ensuring that the gain and loss of fluid is adjusted to meet the body's needs. Correct balance is essential for normal cell function and efficient metabolism. The *intake of fluid* comes from food (approximately 0.75 litres) and, more significantly, from drinking (approximately 1.50 litres).

Fluid not taken orally (by mouth), such as parenteral fluid, must also be accounted for in the fluid balance equation. This may have been administered as intravenous infusion or, more rarely, subcutaneously or into the peritoneal cavity.

The main fluid loss takes place by three routes:
- through *the skin* in the form of sweat (approximately 0.75 litres per day in temperate climates);

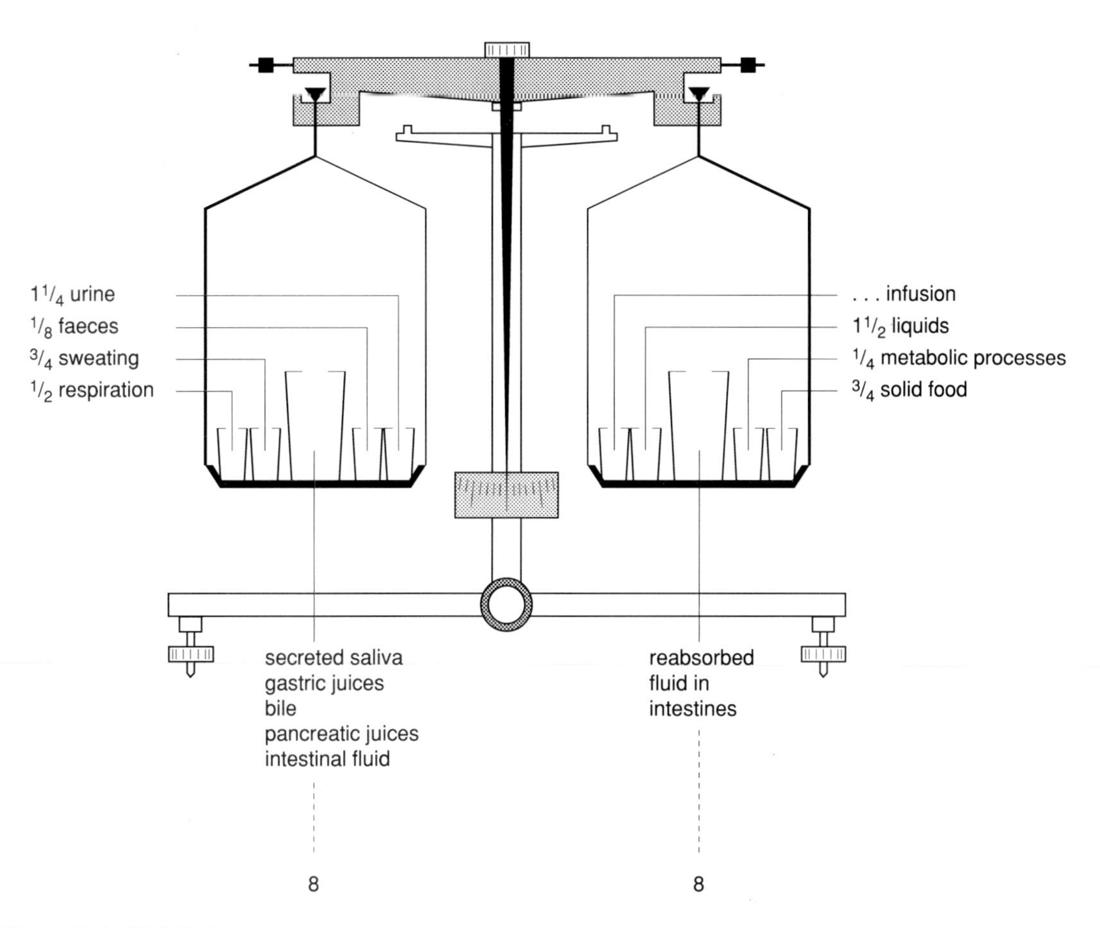

Figure 12.1 Fluid balance
In this diagram of fluid balance the small glasses represent the external intake and loss of fluid, as well as the water production by metabolic processes in body cells. The larger glasses represent the greater volumes of fluid secreted in the intestines, but reabsorbed lower in the intestines in normal conditions. The numbers in the diagram represent volumes in litres. The total intake volume, however, may be subject to considerable variation depending on the composition of daily food (consider the volumes of beer drunk socially in a relatively short period of time by some individuals).

- through *respiration*, from the lungs in breathing (approximately 0.50 litres per day);
- through *the kidneys* in the form of urine (approximately 1.25 litres per day).

This apparent balance can be obviously upset in many common situations such as prolonged thirst, vomiting, diarrhoea, excessive sweating, and excessive renal output.

The kidneys play a prime role in regulating the fluid balance. Normal variations in intake brought about by normal variations in eating and drinking cannot be compensated for by respiratory, sweat, or faecal loss. Regulation of fluid balance therefore requires an efficient circulatory system.

A decreased intake of fluid leads to a decrease in the volume of circulating blood, with a resulting drop in blood supply to the kidneys, and thus in renal function leading to a reduced volume of urine production. Hormonal control modifies renal activity. The specific endocrine actions are brought about by the antidiuretic hormone and aldosterone.

Antidiuretic hormone (ADH) is produced in the posterior lobe of the pituitary. ADH suppresses the secretion of fluid in the renal tubules. *Aldosterone* is produced in the adrenal cortex and functions by preventing excretion of potassium into the renal tubules. Converse mechanisms control fluid balance if excess water and or electrolytes are taken in. The reaction of ADH is more rapid than that of aldosterone.

These regulatory mechanisms have limits of physiological compensation in maintaining homeostasis. If the production of urine is lower than 25ml per hour or 300 ml per 24 hours the condition is defined as *oliguria*, and renal function is endangered. *Anuria* is defined as urine production of less than 10ml per hour or 100ml per 24 hours. Anuria leads to a build up in levels of toxic waste products, cell damage, and resulting uraemia and acidosis.

Figure 12.2
Patient receiving an intravenous infusion

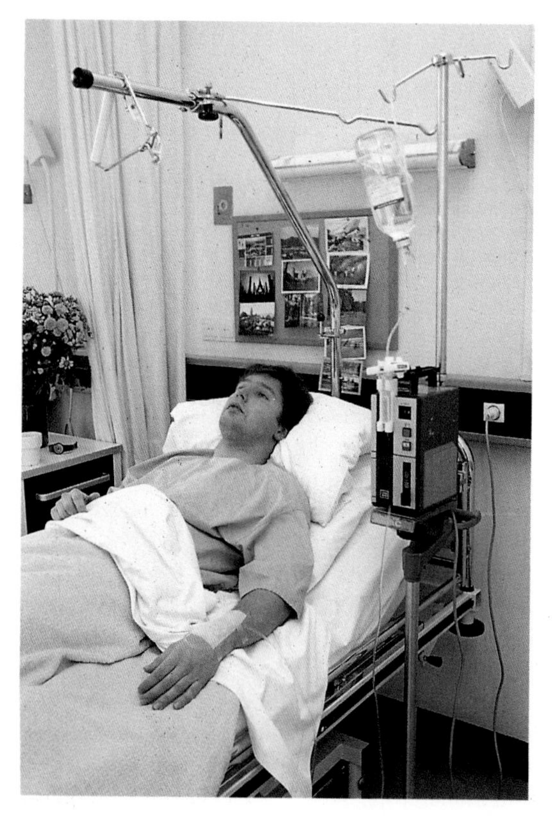

By contrast, excessive administration of fluids and electrolytes, especially by intravenous infusion, will lead to fluid overload.

The careful recording and maintenance of normal fluid/electrolyte balance is clinically essential (Figure 12.2). Loss of fluids and electrolytes due to vomiting, polyuria, diarrhoea, severe burns, major surgical procedures, or through catheters and drains must be measured exactly and appropriately replaced by oral or intravenous (parenteral) administration. In circulatory overload diuretics have a therapeutic role.

Major surgical procedures of any type lead to upset metabolism and temporary catabolism, with reduced function of cell wall fluid transport (the so called'sodium pump'). This needs careful monitoring to avoid overinfusion as eventually normal cell wall mechanisms recover and fluid which was apparently lost is reabsorbed. After abdominal surgery, temporary intestinal paralysis (paralytic ileus) is not uncommon, and this again leads to fluid balance mechanism dysfunction. In such cases, special care must be taken with potassium levels.

3. Summary

The composition of fluids and electrolytes within cells (the internal water balance) is strictly regulated. The overall external water balance in the body controls the internal water/ electrolyte balance by the mechanisms which have been described. The balance between intake and output of fluid has been shown, and the major importance of the kidneys in regulating the fluid/electrolyte balance has been explained.
Disturbance of the normal renal function, with inevitable disturbance of the fluid/electrolyte balance, will result in serious illness. The importance of accurate recording of fluid/ electrolyte balance cannot be overstressed.

Shock

13

1. Introduction

In an attempt to understand some of the effects of shock on the body, a comparison with a domestic central heating system is useful. A loss of pressure in the system results in poorly heated or cold radiators. Similarly, when the pressure in the circulatory system falls below normal levels there is insufficient blood supply to the tissues, and this can have serious consequences. In its effort to prevent these repercussions the body adopts homeostatic mechanisms to correct the changes and in particular to maintain circulation to the vital organs. A patient in a state of clinical shock is in a potentially serious situation and rapid diagnosis and adequate treatment is of vital importance.

Learning outcomes

After studying this chapter, the student should be able to;
- explain the general concept of shock;
- understand the processes of reversible shock;
- understand the processes of irreversible shock;
- identify four causes of shock, and the types of shock they are associated with;
- list six symptoms of shock;
- explain the general principles of the treatment of shock.

2. Pathophysiology

Blood in the circulatory system provides the direct connection between the cells of the various body systems. The blood supplies the cells with necessary nutrients and carries away their waste products. A disturbance of this circulation is therefore an immediate threat to each cell in the body. Insufficient volume of blood or decreased flow in the vascular beds constitutes the existence of shock, which inevitably and immediately causes severe cellular damage. In most cases, it is possible to maintain sufficient circulation by means of a number of compensatory mechanisms to enable the body to recover from any transient damage suffered. In such cases the term *reversible shock* is used. Sometimes, however, these compensatory mechanisms fail, or the duration of shock is too long. In these cases the term *irreversible shock* is used, and massive cellular death is then inevitably followed by the demise of the patient. In every condition of shock a system of regulatory mechanisms, largely controlled by the central nervous system and by a number of hormones, more or less guarantees that the supply of blood to the major vital organs such as brain, heart, kidneys, and lungs, is maintained for as long as possible.

Firstly, the supply of blood to the skin muscles and digestive tract decreases, since these body systems are better able to withstand a temporary reduction in blood supply. The circulation in shock is thus characterised by a well supplied 'kernel' (brain, heart, kidneys, lungs) and a relatively poorly supplied 'peel' (skin, muscles, digestive tract). This partitioning of the circulation is a protective survival mechanism. Insufficient cardiac circulation would lead to pump failure and compound the state of shock as the vital pumping action weakened, and effective output might eventually cease altogether. The kidneys react to an insufficient circulation with an abrupt decrease in waste products, including acid waste. After renal function has ceased, the patient will die from uraemia and acidosis. Insufficient blood supply to the liver leads to a decrease of normal hepatic activity both in terms of manufacturing ability and metabolism of potentially toxic metabolic by-products. The classic signs and symptoms of liver failure will ensue. The brain is especially vulnerable in shock. Loss of an adequate oxygen supply to the brain for only a very few minutes leads rapidly to cerebral cell death. Loss of cortical brain function will rapidly be followed by 'brain death'. The brain is extremely sensitive to any decrease in blood supply and becomes unable to control the regulation of body system functions as loss of neural transmissions occurs. A poor supply of blood to the lungs leads to a shortage of indispensable oxygen and an accumulation of carbon dioxide. Loss of adequate circulation, both in terms of blood perfusion and oxygen supply, to these vital and interdependent organs will therefore, if it is more than transient, inevitably and rapidly prove fatal.

Shock may, then, be described as a serious and life-threatening deregulation of the circulation, which directly depends on four factors to function effectively:

- the pumping force of the heart muscle;
- the total volume of available blood;
- the size and efficiency of the vascular system;
- the functioning of the kidneys.

From these factors, four causes of shock can be identified (Figure 13.1).

1. the cardiac pump fails: *cardiac shock*
2. the circulatory system is insufficiently filled: *low-volume* or *hypovolaemic shock*
3. the vascular system volume is too large due to vasodilation: *vasodilation shock*. This can be divided into two types called *anaphylactic shock* and *septic shock*
4. an abnormal accumulation of blood volume in a closed compartment in the vascular system leaves too little blood available for normal perfusion: *distribution shock*.

Cardiac shock may occur after a myocardial infarction or after other cardiac muscle disorders such as fibrosis in ischaemic heart disease. Heart valve disease (either stenosis or incompetence) can also result in cardiac shock. Less commonly, a pericardial effusion (accumulation of fluid in the pericardial sac) can cause cardiac shock. In cardiac shock the heart muscle, especially the left ventricle, fails to pump sufficient blood volume, leading to blood accumulation 'downstream' in the pulmonary venous circulation. Eventually, the fluid overload in the lungs leads to fluid being forced into the alveoli. The patient will show the signs of'air hunger'and will appear cyanosed (bluish) through lack of systemic oxygen perfusion. In more severe cases pink frothy fluid wells into the upper airways and is expectorated (pulmonary oedema). In cardiac muscle failure of the right ventricle, the systemic circulation becomes fluid overloaded. Such cases show swollen jugular veins, congestion and enlargement of the liver, and peripheral oedema especially of the lower extremities and the soft tissues of the lower back (sacral oedema).

Low-volume or hypovolaemic shock

Blood or fluid loss can occur in many ways. Sometimes the blood loss is visible and can be measured, at other times it is occult and not readily measurable. Although the effects on the circulation and the potential for shock are similar in either instance, the clinical problems

Figure 13.1.
Causes of shock

1. *inferior vena cava*
2. *liver*
3. *portal vein*
4. *capillaries*
5. *aortic arch*
6. *pulmonary artery*
7. *heart*
8. *peripheral vessels*
9. *spleen*
10. *aorta*
11. *stomach and intestines*

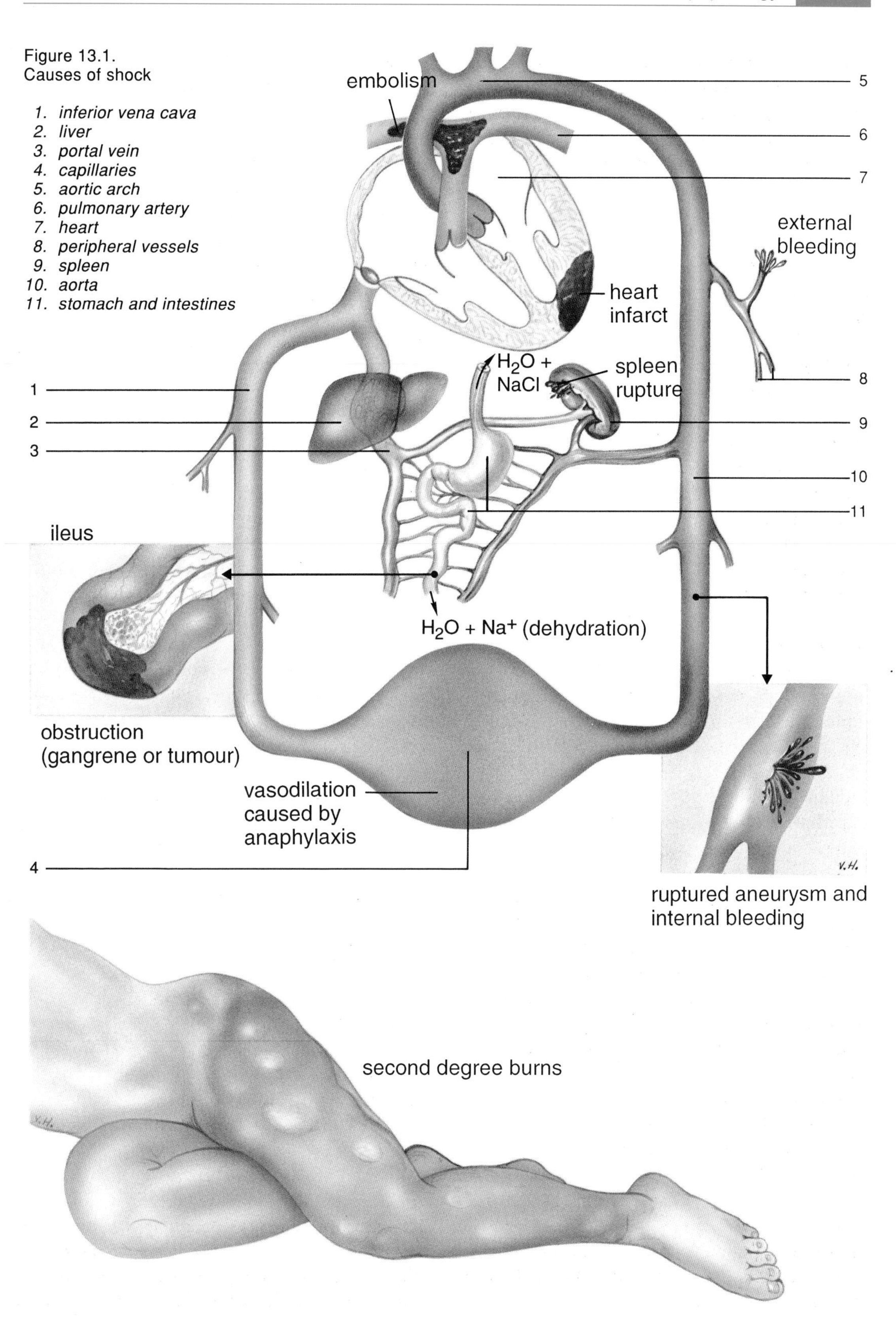

involved in detecting the latter can be especially difficult. External loss of blood is found in numerous conditions, including a bleeding traumatic wound or the vomiting of blood (haematemesis) from a gastric or duodenal ulcer or neoplasm, or from an oesophageal lesion such as a tumour or oesophageal varices.

Blood loss into the intestines is manifest later as darkened faeces (melaena). When vomiting of blood is absent and the patient shows no evidence of melaena it may be difficult to establish a diagnosis.

Expectoration of blood (haemoptysis) due to respiratory tract disorders and *haematuria* (blood in the urine from the kidneys or bladder) provide a good clinical guide to the site of blood loss. Internal blood loss may be more obscure in cases such as rupture of liver or spleen (often traumatic in origin), rupture of the aorta from an aneurysm, a ruptured ectopic pregnancy or a fracture of a long bone such as the femur.

Almost as often, loss of fluid other than blood is a causative factor of shock. In extensive burns, peritonitis, certain kidney disorders (nephrotic syndromes) and in some types of enteritis, large amounts of fluid and protein are lost. The protein loss attracts further fluid loss by osmosis, and administration of plasma or plasma-substitutes may be necessary. In vomiting, some types of diarrhoea, and excessive perspiration, a large volume of fluid which is low in protein but rich in electrolytes is lost. This is also true in polyuria (excessive secretion of urine) in diabetics. Glycosuria (glucose in the urine) also attracts fluid loss by osmosis. Hypercalcaemia in parathyroid disorders and sodium loss in Addison's disease (an adrenal cortical disorder) similarly lead to polyuria.

Fluid loss in paralytic ileus is difficult to quantify and is often underestimated. Large volumes of fluid in the form of gastric, pancreatic, biliary and intestinal juices remain abnormally in the intestinal tract and do not re-enter the circulation as usual while intestinal function is paralysed, and the circulation therefore suffers from a temporary loss of volume; this can lead to shock.

In general, treatment must include administration of fluid, often intravenously, of a composition similar to that lost, together with treatment of the underlying cause.

Vasodilation shock
Septic shock is one form of vasodilation shock. In some types of septicaemia, bacterial pyrogens and endotoxins paralyse the arteriole sphincters. This results in vasodilation of the vascular beds, especially those in the skin and digestive tract. As a direct consequence, the overall circulatory rate and tissue perfusion drops dramatically. This is endotoxic shock. The condition is difficult to recognise because the patients are not pale and do not feel cold. Immediate treatment with plasma or plasma substitutes together with appropriate antibiotics and sometimes corticosteroids and vasopressors may be of vital importance.

Another type of vasodilation shock is anaphylaxis, which is a severe allergic reaction. In anaphylactic shock, sudden dilation of vessels again occurs, caused by certain substances released as part of an abnormal hypersensitivity reaction. Treatment may require speedy administration of adrenaline and sometimes systemic steroids.

Distribution shock
This type of shock occurs mostly in relation to gastrointestinal disorders such as peritonitis, gastroenteritis, and some cases of paralytic ileus. Not only does normal peristalsis cease but intestinal transport mechanisms also fail. Blood and other fluids are not recirculated from the abdominal cavity. This local pooling deprives the general circulation of significant volumes of fluids. Pain may also be a contributory factor in the genesis of paralytic ileus. Shock in pulmonary embolism is also partly brought about as a result of impaired distribution. Blood accumulates in the pulmonary circulation due to the occlusion of the pulmonary arteries by thrombo-embolus and thus cannot reach the left ventricle. As a result, the rate and volume in the

systemic circulation becomes suddely insufficient.

3. Signs and symptoms of shock

The signs and symptoms of shock reflect the underlying processes which have caused it, as well as the mechanisms utilised by the body to mitigate and as far as possible correct the changes.

The most common signs and symptoms of shock are (Figure 13.2):
- pallor;
- sunken face;
- perspiration;
- rapid, weak pulse and low blood pressure (hypotension);
- decreased production of urine (oliguria);
- anxiety and at times confusion.

The basis of the oliguria and hypotension have already been described.

Pallor
The decreased volume in the circulation or the impaired rate of flow triggers a compensatory vasoconstriction in specific vascular beds, especially the skin muscles and digestive tract, to maintain adequate supplies to vital centres such as the brain. The skin will become pale and feel cold.

The *sunken face* and deep-set eyes often observed are due to peripheral vasoconstriction. A related phenomenon is slow capillary refilling, which can be seen when after short pressure has been applied to the skin the obvious blanching takes longer than anticipated to recover.

Perspiration is explained by the fall in temperature of the skin as a result of vasoconstriction. The heat sensors in the skin detect that the environment is warmer than the skin and they react by stimulation of perspiration. A cold, moist skin is always a good indication of shock.

The *weak, rapid pulse* is a direct result of the decrease in the volume of blood pumped by the heart, either due to the impaired force of the heartbeat or to the fact that the volume of blood returning to the heart has considerably decreased. This also explains the fall in systolic blood pressure. At times the pulse may be barely palpable.

A *decreased volume of urine* is observed in most types of shock (except those caused directly by polyuria). In shock there is reduced renal blood supply and the reduced kidney perfusion leads to oliguria. This partially acts as a compensatory mechanism to maintain blood volume which would otherwise be reduced.

The *conscious state* of the patient is often striking. The patient, although anxious, is often self-absorbed with little awareness of the external environment. Elderly patients often react with mental dulling and may be confused and markedly restless.

4. Treatment

Treatment must include general and specific measures.

Generally, the patient in shock is extremely vulnerable.
Excessive transportation and movement, unnecessary examinations, and superfluous diagnostic procedures should be avoided. The outcome is much improved if the patient is nursed in a supine position with the lower limbs slightly raised if possible, and kept as quiet and relaxed as possible in a warm (not hot) room without unnecessary noise or disturbance. Administration of oxygen is needed in most cases. If vomiting presents problems a naso-gastric drainage tube may help. The body temperature, pulse, respiration, blood pressure, and central venous pressure must be regularly monitored.

Accurate recording of the fluid balance (oral and parenteral fluid, urine, faeces, vomitus, catheter and drainage fluid) is an absolute necessity. An incontinent or unconscious patient may require bladder catheterisation.

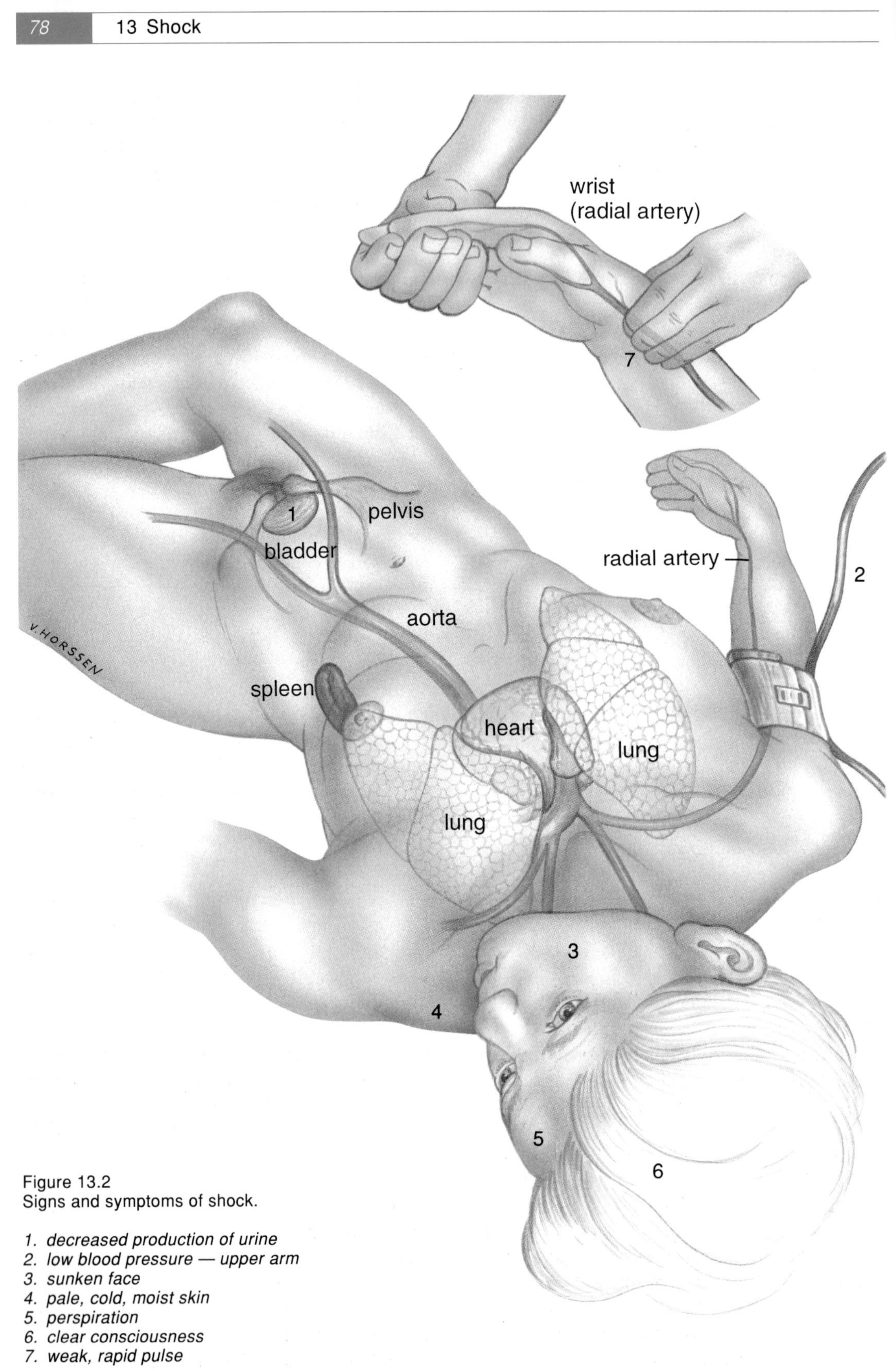

Figure 13.2
Signs and symptoms of shock.

1. *decreased production of urine*
2. *low blood pressure — upper arm*
3. *sunken face*
4. *pale, cold, moist skin*
5. *perspiration*
6. *clear consciousness*
7. *weak, rapid pulse*

Such detailed care is often carried out in a dedicated 'intensive-care' unit where ECG monitors, telemetry and respiratory ventilators are more readily available. The specific methods of treatment are determined by the specific causative factors.

In general terms, resuscitation should be undertaken prior to intervention surgery, though obviously in cases of rapid and life-threatening blood loss this may not be wholly possible. In cardiac shock, vasopressors may be required. Pulmonary causes require ventilation. Antibiotics will be necessary in cases of septic shock, as will appropriate doses of insulin in diabetics. Severe shock in Addison's disease will need corticosteroids, and in renal failure haemodialysis may be undertaken.

Rapid fluid-loss and blood-loss replacement requires intravenous administration. Whereas most cases can be managed by peripheral vein approaches, a long line to a more centrally situated vein (often the subclavian vein) may sometimes be required. Infusion flow rates can now be very accurately controlled by electronic monitored pumps. When oxygen carriage is required without the need for total blood fluid volume, packed cells (blood with much of the plasma removed) are used.

In all cases of blood transfusion careful cross matching of the donated blood will be undertaken. Nevertheless, all patients receiving blood need careful observation to ensure that any adverse reaction, which would in itself produce shock, is caught early. Mismatch with associated red cell haemolysis can itself be fatal. Blood transfusion laboratories always have meticulous regimes and protocols to minimise blood mismatch, and all patients receiving blood must be precisely matched against the details on the blood transfusion product. This information must be checked and rechecked before the product is administered, and all the details must be carefully recorded. Sometimes, despite all the cross-matching checks being correctly carried out, the patient receiving a blood transfusion will experience an elevated temperature and may start to shiver or rigor. This may be as a result of obscure leucocyte incompatibility. Previous history of such reactions may alert the clinician and (if time and the clinical condition permit) allow the administration of blood with most of the leucocytes removed.

Blood transfusion centres test all blood donated for hepatitis B, HIV, and other infectious agents, and careful selection of potential donors reduces the risk of HIV in blood supplies. (Previous disasters with HIV-contaminated Factor VIII given to haemophiliacs are well known.) Blood products are also now available as specific sub-components. 'Packed' red cells have most of the plasma removed to avoid potential fluid overload and congestive cardiac failure. Leucocyte suspensions are available for patients with agranulocytosis. Platelet suspensions are used in thrombocytopaenia. Factor VIII is available for haemophiliacs. Some specialised centres have other specific clotting factor concentrates.

Plasma and individual plasma proteins can be given where protein depletion is a problem. In severe protein depletion albumin may be administered to correct the colloid osmotic pressure, otherwise a considerable volume of fluid leaves the circulation and causes oedema.

Specific globulins are used in immuno-deficiency conditions and therapeutically or prophylactically for certain infections (such as the prevention of hepatitis A for international travellers).

It is worth noting that in cases of catastrophic haemorrhage, such as severe trauma or ruptured aortic aneurysm, the total normal volume of the patient's blood may be replaced and the patient may survive.

Apart from blood products, parenteral administration of electrolytes including sodium, potassium and calcium are required, as well as glucose. Replacement of electrolytes and glucose are needed to maintain the correct extracellular/intracellular balance. In normal physiological situations sodium ions are predominantly extracellular and potassium ions intracellular, and a precisely regulated 'sodium pump' mechanism exists to maintain the correct balance on each side of the cell walls.

When delivering intravenous fluids it is essential that they are in an isotonic state. This means that the concentration of glucose and electrolytes is the same as that in the cells, and consequently the same as that in the red blood cells. Should the concentration be too low (hypotonic) there is too much water in the solution and the red blood cells will absorb water, swell, and burst (haemolyse). A too-concentrated solution (hypertonic) causes cellular dehydration. Both complications obviously must be avoided.

Commonly used isotonic solutions are:
- *sodium chloride solution* (0.9% NaCl). This contains 9 grammes of sodium chloride per 1 litre of solution. In specific cases a hypotonic solution of 0.45% NaCl is available;
- *Ringer solution*, containing sodium chloride, potassium chloride and calcium chloride;
- *sodium bicarbonate solution* (1.4% NaHCO3). This is used to correct acidosis;
- *glucose solution* (5% glucose). This also provides 837 kJ energy.

Solutions are prepacked and sterile and in practical volumes usually of 500 ml.

Prolonged administration of isotonic solutions can lead to a deficiency of many types of nutrients and to insufficient calorific intake. As a result the metabolism of the cells will not be able to meet demands. Repair and reconstruction at a cellular level will be in negative balance as the blood and plasma lack the basic building materials. To meet these problems supplementary nutritional replacement fluids are available. Administration of a solution with 5% glucose is insufficient in providing energy needs in acceptable volumes. To meet energy needs the volumes would be excessive and could lead to fluid overload, congestive cardiac failure and pulmonary oedema. The following are more suitable:
- 10% or 20% *glucose*, invertase, and *sorbitol* are available for high calorie replacement. Such solutions, however, are not without problems and may cause phlebitis (venous inflammation);
- 5% or 10% *aminoacetic acid solutions* can provide amino acids as building blocks for proteins to aid cellular repair as well as other essential metabolic requirements;
- *lipids*, in the form of emulsions, provide not only fatty acid products but also high calorific value per unit volume. Lipid solutions need careful monitoring to avoid overload and emulsion separation which has a micro-embolic potential.

5. Summary

Loss of circulation and perfusion to vital organs rapidly leads to progressive and permanent cell damage and cell death. Lack of rapid correction will lead to death of the patient.

The type of shock depends on the causative conditions.
Shock can be classified as:
- cardiac;
- hypovolaemic;
- vasodilation (septic or anaphylactic);
- distributive.

The signs and symptoms reflect the type of shock and the underlying condition. Treatment is directed at replacement and general somatic resuscitation together with management of the underlying causative disorder.

Oedema

14

1. Introduction

In this chapter, *oedema* refers to an excessive accumulation of fluid in the tissues. This is a result of disturbed exchange of fluids of the tissue capillaries – the old lay term, sometimes still heard, is 'dropsy'. *Lymphoedema* is excessive accumulation of fluid in the lymphatic system. *Myxoedema* is marked by an abnormal deposition of muco-protein beneath the skin, and this is found in some cases of hypothyroidism.

Learning outcomes

After studying this chapter, the student should be able to:
- explain the exchange of fluid at capillary level;
- define the term oedema;
- describe the four main causes of oedema;
- identify the differences in location of oedema in a patient while upright and when supine.

2. Fluid exchange in the tissues

There is a continuous fall in blood pressure in the circulation towards the periphery. The initially high pulsatile pressure wave in the aorta and major arteries, caused by the contraction of the left ventricle of the heart, gradually and slowly diminishes as the surface area of the peripheral vascular bed increases at arteriole level. At the junction between the arterioles and the arterial part of the capillary, the pressure has dropped to 30 – 35 cm of water. The pressure falls even further during the passage through the capillaries until it is only 10 – 15 cm of water at the venous end of the capillaries, where the blood flows into the venules. These peripheral pressures are referred to as *hydrostatic pressures*.

As the capillaries have thin walls without muscular layers it is apparent that even at relatively low pressures large volumes of fluid can easily be discharged into the extracellular spaces. This volume has been estimated to be as large as 2000 litres per day. However, by normal homeostatic mechanisms a comparable volume is returned to the circulation. This is achieved by two main mechanisms.

Tissue pressure is the first mechanism. This arises from the pressure of fluid already present in extracellular spaces. Thus, the more fluid already present, the higher the pressure.

The second mechanism is *colloid osmotic pressure*. This is created by the presence of proteins, especially albumin, in the blood.

These large molecular proteins cannot permeate the vascular walls in normal conditions, and they thus create a differential concentration across the capillary wall. This differential concentration generates a colloid osmotic pressure forcing fluid back into the vascular space. As already stated, at the arterial side of the capillaries the hydrostatic pressure is between 30 and 35 cm of water, which is somewhat higher than the sum of the colloid osmotic pressure and the tissue pressure. The albumin at this site is relatively diluted by unit blood volume and the tissues are relatively poor in fluid, hence low tissue pressure. As a result a considerable volume of fluid leaves the capillary at the arterial end.

At the venous end of the capillary the pressure ratios are reversed. The hydrostatic pressure has dropped to 10 – 15 cm of water, whereas the tissue pressure has risen. The colloid osmotic pressure has also risen as the concentration of albumin has increased per unit volume as a result of previous fluid loss.

In the process, fluid rich in oxygen and soluble nutrients, pushed into the tissues at the arterial end of the capillaries, becomes available to needy cells. All these substances having been transferred, fluid returns to the venous end of the capillaries carrying the carbon dioxide and waste products resulting from metabolism. Circulatory homeostasis is therefore maintained, and the cells are supplied with oxygen and nutrients. The efficient removal of carbon dioxide and waste products also takes place.

3. Causes of oedema

Oedema exists when there is an excessive accumulation of fluid in the tissues as a result of disturbances in the exchange of fluids of the tissue capillaries. Consequently, oedema can only occur when fluid leaves the bloodstream at the arterial end of the capillary and does not return at the venous end. One obvious cause is too high a pressure in the venous portion of the capillary, especially if the venous flow is insufficient or obstructed. In such conditions there is said to be local venous stasis. This may be caused by a local disorder such as venous thrombosis, or it may arise as a result of local pressure on veins caused by, for examples, an overdistended bladder, a pregnant uterus, an enlarged prostate, a uterine tumour, or a colonic mass, all of which result in an abnormal pelvic swelling.

Another common situation is that the heart is not forceful enough to pump away all the venous return due to impaired pumping ability of the right ventricle (right heart failure). This is often seen in chronic obstructive airways disease. An obstruction due to a mediastinal neoplasm or enlarged mediastinal lymph nodes can also impede venous return in the great veins, especially the vena cava.

A totally different scenario exists if there is a marked decrease in colloid osmotic pressure. Colloid osmotic pressure is the most important factor in providing suction power to bring about the re-entry of tissue fluid into the venous end of the capillaries. As this pressure is directly proportional to the concentration of protein in the blood, any fall in protein concentration, especially albumin (*hypoalbuminaemia*), is inevitably accompanied by oedema. A fall in albumin is found in some renal diseases when there is albumin loss through the urine (*albuminuria*). Some intestinal disorders lead to protein loss (protein-losing enteropathy), and in liver conditions, such as hepatic cirrhosis, too little protein is produced. Protein insufficiency may also be found in cases of undernourishment such as kwashiorkor in developing countries, and in starvation, especially in the elderly if in a neglected condition.

Other renal conditions lead to fluid absorbtion but insufficient excretion: here the tissue fluid pressure increases but the colloid osmotic pressure decreases due to dilution. The pressure at the venous end of the capillaries then rises and oedema develops. Clinically, in cases of rapid onset of renal disease such as acute glomerulonephritis, the oedema is first observed around the eyes.
Another situation in which oedema (especially

local oedema) occurs, is when a local inflammatory condition exists. The inflammatory process causes microscopic damage to the vascular wall, with local impaired function. The walls become permeable to proteins causing the colloid osmotic pressure to fall locally almost to zero. Examples of this are seen surrounding abscesses, in hypersensitivity reactions such as urticaria and oedema glottitis, and in tissues which have been burned.

4. Signs and symptoms of oedema

Latent oedema exists when fluid accumulates in the tissues but is not visible. It is only detected by an increase in the patient's weight. Sometimes the fluid collected in the tissues during the day is excreted at night, leading to *nocturia*. The site where oedema forms is mainly determined by body position, as the oedematous fluid accumulates at the lowest part of the body under the influence of gravity. In the sitting or standing position, the feet and ankles are the first to swell and the patient notices that shoes are tight or no longer fit. The patient may in fact notice a groove forming above the top of the shoes due to pressure. Clinically, oedema can be demonstrated in the legs by applying local digital pressure, leaving a marked local depression. This is best demonstrated where there is bone closely underlying, such as the inside of the ankle or the front of the lower leg (Figure 14.1).

After lying in the supine position for some time, the fluid gravitates to the lower back and the oedema is detectable over the sacrum and lumbar vertebrae (Figure 14.2). Bed sheets may

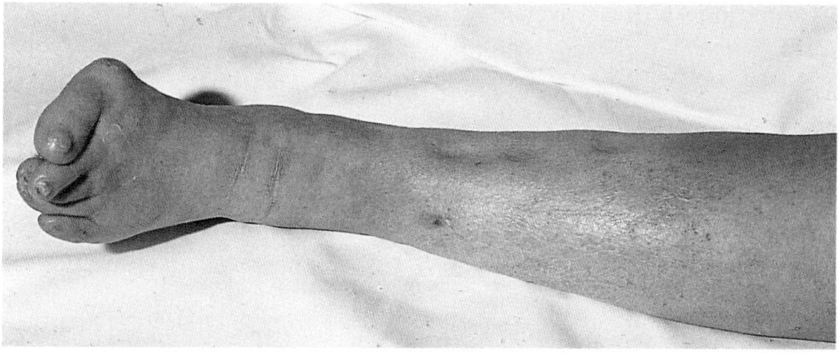

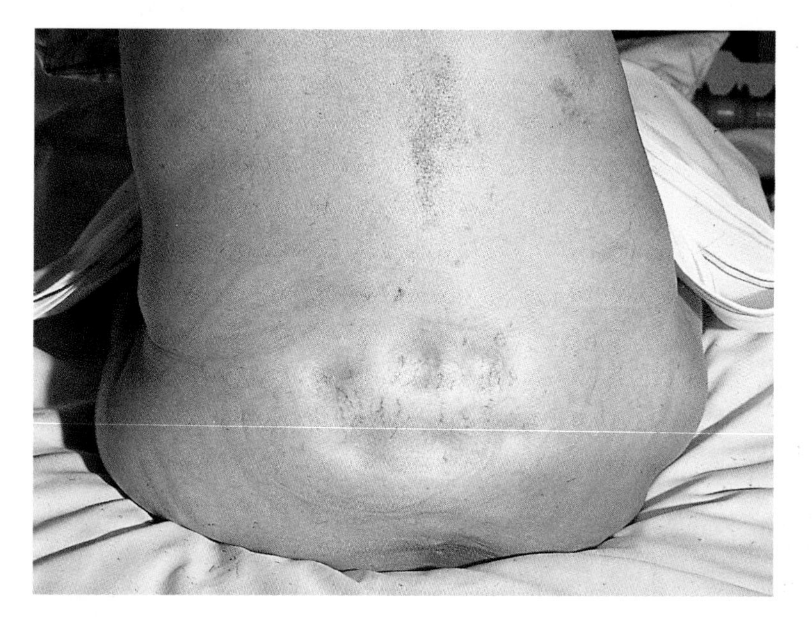

Figure 14.1
Oedema of the lower leg

Figure 14.2
Sacral oedema

also leave noticeable creases on the patient's skin.

Oedema may also accumulate at the inner aspects of the thighs, in the scrotum in males, and in the vulva in females. Generally, the more serious the oedema the higher up the body it is observed, reaching in very serious cases to the armpits (axillary oedema). Oedema may accumulate in the peritoneal cavity forming ascites or in the pleural cavity forming one type of pleural effusion.

5. Summary

The pressure balances in the circulatory system are very carefully controlled and monitored, so oedema does not occur in the healthy individual. But when abnormalites arise in the concentration of blood protein or in the venous pressure, or where there is local inflammation, systemic or local oedema formation will develop.

Respiration

15

1. Introduction

The organs of the respiratory tract carry air from outside the body to the sites where gases can be effectively exchanged. There is then a two-way process consisting of oxygen intake and carbon dioxide and water vapour expulsion. The mechanical transport of air into and out of the lungs, and passage of gases into and out of the blood in these organs is called *external respiration*. The subsequent passage of gases into and out of cells and tissues around the body is called *internal respiration*. This chapter is primarily concerned with external respiration.

The respiratory tract consists of the oral and nasal cavities, the larynx, the trachea, major and minor bronchi, small bronchioles in the lungs, and finally the alveoli where the gaseous exchange takes place.

Learning outcomes

After studying this chapter, the student should have sufficient knowledge and understanding of:
- the principles of respiration;
- the physiology of the different types of respiration;
- respiratory disorders;
- auxiliary respiration;
- Cheyne-Stokes and Kussmaul's respiration and hyperventilation.

2. Function of respiration

The ultimate function of the respiratory tract is the exchange of gases.
This takes place in the alveoli, where oxygen from the outside environment is absorbed by the blood for conveyance to the tissues which require it for normal metabolic processes. The tissues release carbon dioxide which is absorbed in the blood and carried to the alveoli where it is transferred back to the atmospheric air. Thus, the expired air obviously contains less oxygen and more carbon dioxide than the inspired air. Intake of air is brought about by suction during inspiration (or inhalation). During inspiration the space in the thoracic cavity increases as the ribs rise and the diaphragm muscle flattens. Expiration (or exhalation) takes place by passive elastic mechanisms.The muscles relax and because of the elasticity of the pulmonary tissues and the increased pressure in the abdominal cavity, the ribs descend and the

volume of the thoracic cavity decreases. Pulmonary tissues have an intrinsic elastic property and a tendency to contract. A combination of anatomy and the laws of physics dictates that there is normally a negative pressure, or partial vacuum, in the thoracic cavity, which means that there is a lower pressure in the thoracic cavity than in the outside environment. As a consequence, the lungs remain in a distended state. The normal respiratory rate is approximately 16 times per minute in adults (it is much higher in infants).

Dead space

The upper parts of the respiratory tract do not participate in the gaseous exchange function but they are nevertheless important for transportation. The volume of this space in the respiratory tract is referred to as the 'dead space'.

Only a proportion of the inhaled volume of air is involved in gaseous exchange, and the same is true for exhaled air. Some of it stays behind in the dead space. The larger the dead space, therefore, the less effective the gaseous exchange. In emergency clinical situations the dead space can be reduced by performing a tracheotomy, in which a surgical opening is made in the trachea just below the larynx. In this situation the dead space is reduced as the nose, mouth, pharynx and larynx are bypassed and do not take part in the transport mechanism.

Normal respiration follows a regular rhythm. It is controlled by respiratory centres in the brain which are sensitive to the concentrations of oxygen and carbon dioxide in the blood. Since the pH (acidity) of the blood is modified by respiration, there are also brain centres which monitor and react to changes in blood pH.
These centres assess and react to the blood acidity, and control the respiration as necessary. The regular respiratory rhythm may be disturbed by peripheral stimuli. *Laughing,* for example, causes the respiration to become jerky and irregular. A *hiccup* brought about by sudden contraction of the diaphragm results in upset respiratory rhythm (the sound is made as a result of a relatively closed glossopharyngeal orifice). *Sneezing* is a swift, forced respiration accompanied by a reflex closing of the nasal passages. *Coughing* is a forceful expiration, with the help of the auxiliary muscles of respiration, with a closed glottis. Coughing can be productive (when mucous material is expectorated) or dry and unproductive as in a ticklish cough. *Sighing* is a deep inspiration and expiration which may be voluntary or involuntary.

3. Mechanisms of respiration

In normal respiration two mechanisms regularly alternate.
They are *inspiration* (or inhalation) and

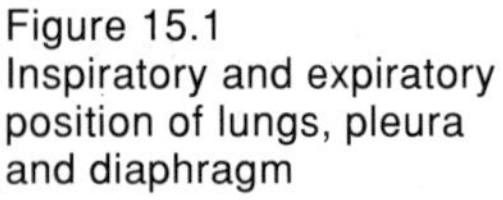

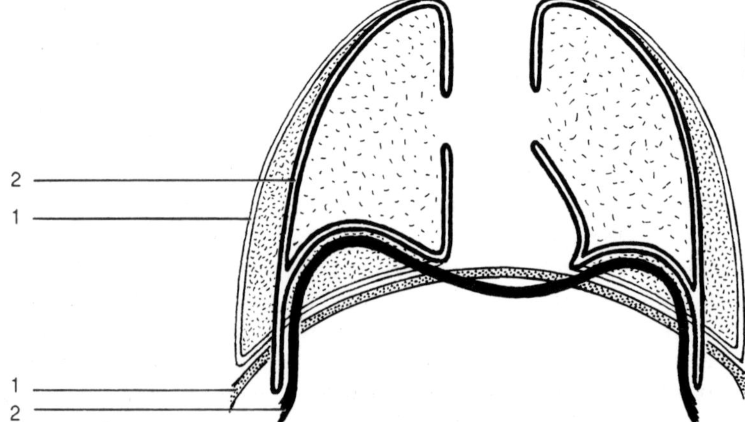

Figure 15.1
Inspiratory and expiratory position of lungs, pleura and diaphragm

1. inspiration
2. expiration

expiration (or exhalation) (Figure 15.1). Inhalation is an active process in which the intercostal muscles move the ribs and chest wall forward and upwards, taking the pleura with them. As there is no air between the parietal and visceral pleura, but only a thin layer of lubricating fluid, surface tension prevails. The visceral pleura is intimately attached to the underlying lungs which must thus follow the pleural movement up and down in concert with the pleura. This constitutes *thoracic* or *intercostal respiration.*

A second mechanism exists. In this process the thoracic cavity expands by contraction of the diaphragm muscle. This is a somewhat unusual dome-shaped muscle with fibres fixed centrally and peripherally, resulting in flattening on contraction. As the diaphragm flattens, the volume of the thoracic cavity considerably increases. The abdominal organs are moved downwards and forwards and the anterior abdominal wall is stretched at each inspiration. This is abdominal respiration. When the abdominal muscles play an unusual or excessive role (as they do in some lung disorders), the term abdominal respiration is applied.

Both thoracic and diaphragmatic movements take place simultaneously. As a result of this double movement the volume of the thoracic cavity increases to a considerable extent. The lungs expand and air, rich in oxygen and poor in carbon dioxide, enters via the nose, trachea, bronchi, and bronchioles to reach the alveoli. The oxygen in the alveoli undergoes gaseous exchange to enter the branches of the pulmonary vessels. Carbon dioxide undergoes gaseous exchange from the vessels into the alveoli and is thence expired. The inspiratory movement is stopped by an impulse to the respiratory centre in the brain's medulla oblongata when the lung fibres are at maximum expansion. Contraction of the intercostal muscles and diaphragm ceases, the muscles relax, and elasticity returns the chest cage to its original shape and volume. Reduction in volume during expiration is slower than the expansion during inspiration, and air poor in oxygen and rich in carbon dioxide is expired. The expiratory phase is totally passive, in contrast to the active inspiratory phase, and it lasts one third longer. Expiration is concluded by the action of a respiratory centre, which is extremely sensitive to carbon dioxide, as soon as the concentration of the carbon dioxide reaches a maximum 'trigger' level. The process then begins again with a new inspiratory phase and repeats itself some sixteen times or so per minute.

4. Disorders of inspiration and expiration

Dyspnoea is the general term used to imply difficulty in breathing. This must be distinguished from various misleading lay descriptions relating to chest oriented symptoms ranging from indigestion to cardiac derived pain.

Dyspnoea is usually sub-classified as *inspiratory dyspnoea*, which implies difficulty in inspiration, or *expiratory dyspnoea* which suggests difficulty in expiration.

Inhaled air may be obstructed by a variety of different disorders. Oedema of the pharynx, larynx, epiglottis, or trachea can create barriers, and external pressure from tumours in the thyroid or enlarged lymph nodes has a similar effect.
An aspirated 'foreign body' may cause a sudden life-threatening respiratory obstruction (Figure 15.2a and b). Sometimes the presence of a marked obstruction will cause the inhaled air to generate a characteristic shrill, harsh sound referred to as *inspiratory stridor*.

In expiratory dyspnoea the disorder lies in the lower respiratory tract. The active, forceful inspiration is comparatively unhampered but the passive expiration is severely obstructed. This occurs where there is accumulation of mucus in the smaller airways, where there is excessive contraction, swelling, or oedema of the small airways, or where there is excess fluid in the alveoli. Clinically, a harsh, high-pitched noise can be heard on expiration, and the expiration phase lasts longer than normal. Sometimes expiration lasts so long that total

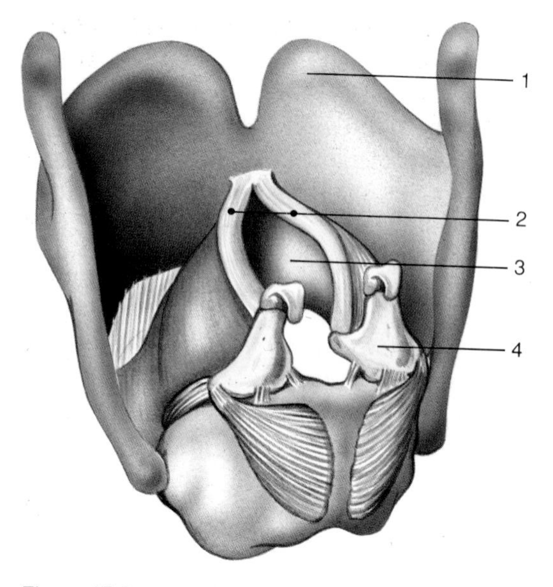

Figure 15.2a
Larynx seen from behind

1. *thyroid cartilage*
2. *vocal cords*
3. *glottis*
4. *arytenoid cartilages*

Figure 15.2b
Larynx with foreign body

1. *root of tongue*
2. *epiglottis*
3. *vocal cords*
4. *side walls*
5. *trachea and glottis*
6. *arytenoid cartilages*
7. *back of larynx*

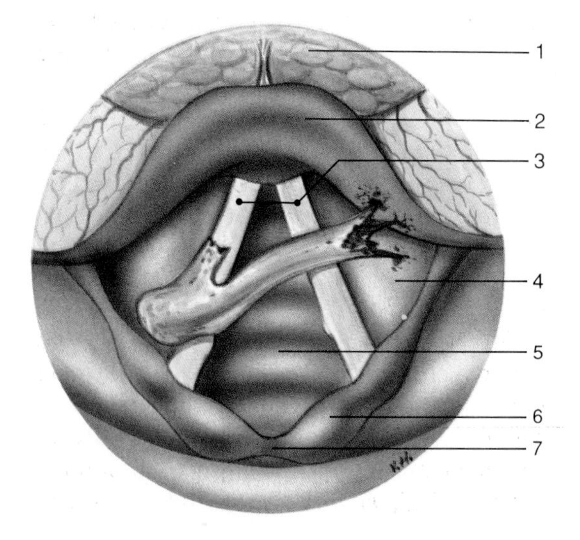

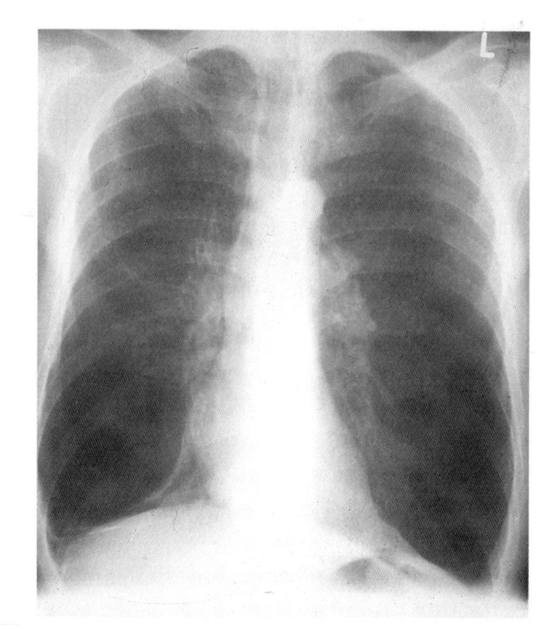

Figure 15.3
Pulmonary emphysema
Note low position of lung contours

expiration never occurs and the following inspiratory process commences before the volume of the chest cavity has returned to normal. As a result, too much expiratory air stays in the airways and lungs. This is seen in cases of emphysema (Figure 15.3).

A dyspnoeic individual will, by reflex, try to improve respiration by using other muscle groups not normally used in the process; the accessory muscles of respiration. These include the muscles between the thoracic cavity and the shoulder normally utilised for shoulder functions.

If the shoulders of the dyspnoeic patient are secured or braced against an object such as a table, bed, or chair, the auxiliary muscle groups raise the chest more effectively and respiration is easier (Figure 15.4). Clinically, a deep groove is observed above the clavicle in cases of auxiliary respiration.

In severely dyspnoeic patients, especially in children, the nostrils are seen to move. This nostril respiration does little to improve matters,

Figure 15.4
Hyperventilation

but it does indicate severe dyspnoea. As noted, the severely dyspnoeic patient often finds some relief by sitting in an upright position. This is particularly notable in cases of congestive cardiac failure with pulmonary congestion and pulmonary oedema.

5. Thoracic and abdominal respiration

In some conditions the thoracic respiration is impaired.
Pain is one major cause. Such pain may arise due to a fractured rib or the presence of a pneumothorax (in which air is present between the two layers of the pleura). Mechanical or neurological lesions may also impede thoracic movement. In these instances the action of the abdominal muscles contributes more than usual to respiration, and the condition is called *abdominal respiration*. Clinically the chest may be seen to be hardly moving, while the abdomen moves in an exaggerated fashion.
Some disease processes, being unilateral, have a corresponding effect on respiration. One important example is pneumothorax, where as a result of trauma to the chest wall or a rupture on the pleural aspect of the lung, air enters the space between the two layers of the pleural membrane and thoracic respiratory movement is seen to decrease. Similar changes are seen when a significant volume of fluid accumulates in the pleural space. Neurological disease can have the same effect. Pain may arise from pleural lesions (*pleurisy*), which in turn are caused by some underlying infection such as pneumonia or pulmonary embolism. Disorders affecting the intercostal nerves can cause severe pain (*intercostal neuralgia*), and trauma to the ribs is also very painful. In all these cases abdominal respiration becomes pronounced. Conversely, severe intra-abdominal disorders, such as peritonitis, immobilise the diaphragm and a compensatory increase in thoracic respiration is found.

6. Changes in depth and rate of respiration

Very rapid respiration is called *tachypnoea*. The terms *bradypnoea*, *hypopnoea* and *hyperpnoea,* referring to slow respiration, superficial respiration, and excessively deep respiration respectively, exist but but these terms are not widely used.

During marked physical exertion the rate and depth of respiration increase, resulting in a so called *dyspnoea of effort*. Comparable changes in respiration are sometimes seen in situations of extreme emotion. If dyspnoea occurs during slight exercise a pulmonary and/or cardiac abnormality probably exists.

In such circumstances too little blood flows to the respiratory centre in the brain, or the blood flowing to it is poor in oxygen. Thus when a patient complains of shortness of breath on normal walking or when climbing a stair or slope, a cardiac or pulmonary lesion should be suspected.

When dyspnoea occurs at rest, the condition is obviously more alarming to the patient and more serious clinically. In extreme cases when the left ventricular output falls dangerously, serious pulmonary congestion and oedema can arise. In this circumstance the patient experiences a choking sensation. This often happens suddenly and at night, giving rise to so called *cardiac asthma*. Copious amounts of fluid are pushed out of the pulmonary vessels into the alveoli (pulmonary oedema). The patient is severely dyspnoeic and starts to wheeze. The skin is pale and clammy and the patient expectorates pink, frothy fluid which is sometimes blood streaked.

Hypopnoea is almost always accompanied by tachypnoea. This is found in severe end-stage pulmonary disorders such as pulmonary bronchitis and emphysema. It is also seen in severe bilateral inflammatory conditions such as bronco-pneumonia. Respiration is not only fast but extremely shallow. The patient is cyanosed and literally gasping for breath. The lips may adopt a fish mouth appearance. Biochemically there is too little oxygen and too much carbon dioxide in the blood. This clinical picture may also be seen in severe trauma, especially crush injuries, shock and drowning, and in 'adult respiratory distress syndrome' from other causes. These conditions may require artificial ventilation.

Cheyne-Stokes respiration is a well-recognised disorder of respiratory rhythm (Figure 15.5). There is initially deep, rapid respiration often accompanied by marked restlessness. Next follows relatively slow and shallow respiration followed by a short period of apnoea when breathing stops and the level of consciousness may fall. After this period of apnoea, slow, shallow breathing changes to rapid, deep breathing. The process then repeats itself, and the whole cycle may take several minutes to complete. The cause is disordered function of the respiratory centre in the medulla oblongata, which may be be caused by local cerebral pathology such as thrombosis, haemorrhage, tumour, or inflammation. Toxic states arising as a result of severe liver damage or due to poisoning from sedatives, painkillers, or neuroleptics have a similar effect. Dying patients often exhibit this breathing pattern as a terminal event.

A less marked but similar phenomenon is found physiologically in some elderly and very young individuals.

Kussmaul respiration (Figure 15.6) is another common disorder of respiratory activity. This consists of continuous, rapid, and noisy respiration in which the respiratory system, as part of the homeostatic mechanism, is attempting to compensate for a metabolic acidosis disorder. It is seen in some cases of

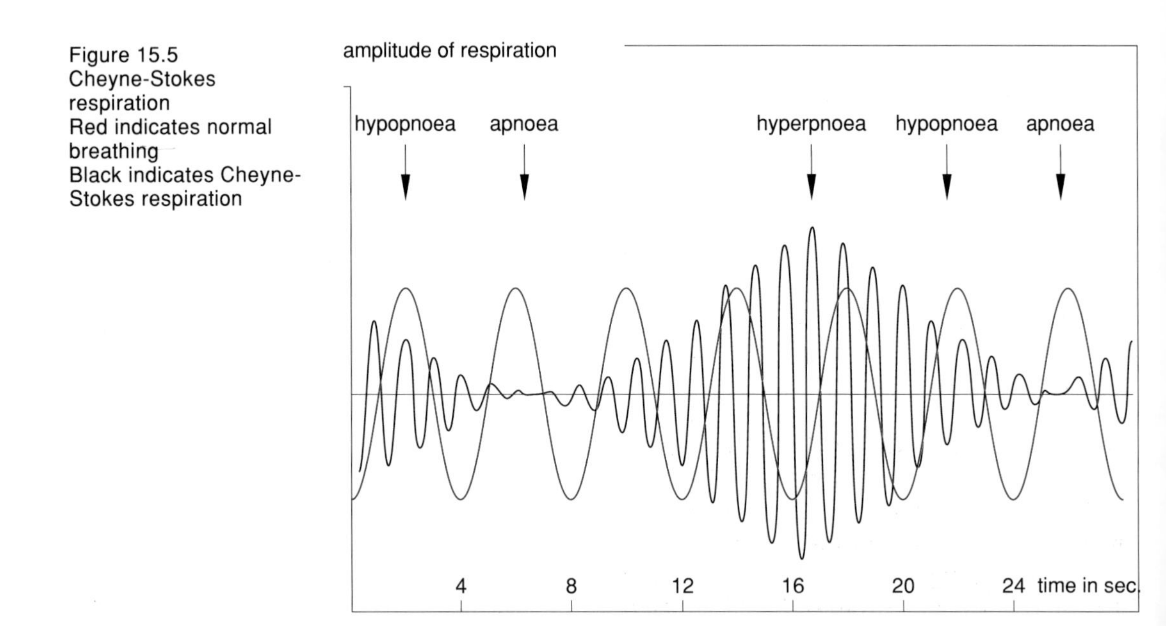

Figure 15.5
Cheyne-Stokes respiration
Red indicates normal breathing
Black indicates Cheyne-Stokes respiration

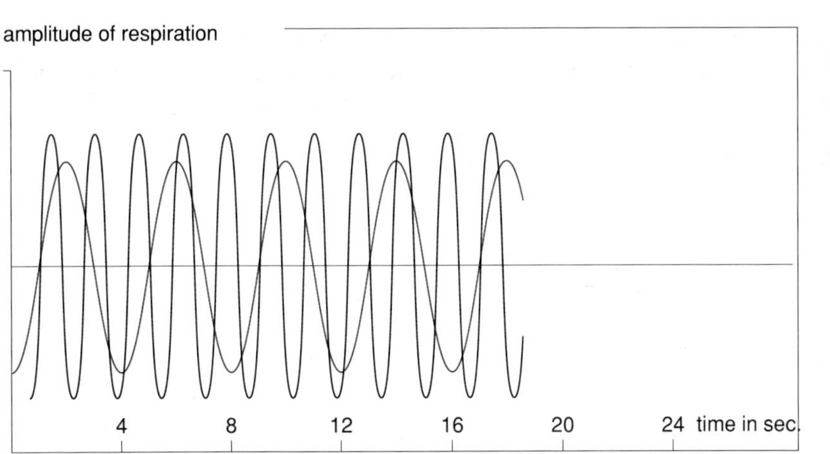

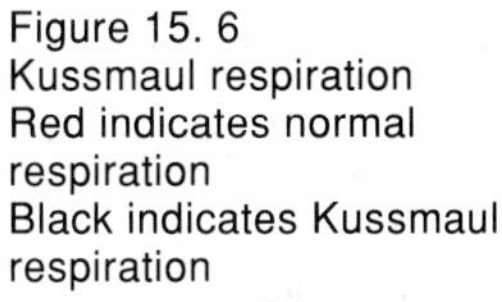
Figure 15. 6
Kussmaul respiration
Red indicates normal respiration
Black indicates Kussmaul respiration

severe renal or hepatic failure and in diabetic acidosis. Respiratory compensation for the metabolic acidosis by increased expiration of carbon dioxide is likely to be only partial and can be evaluated by blood gas analysis.

Hyperventilation may occur in cases of extreme emotion or hysteria. In contrast to Kussmaul respiration, this is not a compensatory mechanism. As a result of hyperventilation, excess oxygen is inhaled and too much carbon dioxide exhaled. The patient experiences a prickling sensation in the arms, hands, and around the mouth, together with tightness in the chest. Spasms of the arms and hands are not uncommon. Treatment by inducing the patient to rebreathe into a closed container such as a plastic bag is rapidly and totally effective.

7. Summary

Normal respiration takes place in two phases; active inspiration and passive expiration. The thorax expands by contraction of the intercostal muscles, giving thoracic respiration, and by contraction of the diaphragm, giving abdominal respiration.

Inspiratory, expiratory, thoracic, and abdominal types of dyspnoea are recognised. Most respiratory disturbances are caused by pulmonary or cardiac disorders. In acidosis Kussmaul respiration is observed, while Cheyne-Stokes respiration is found in both brain and systemic disorders. Cheyne-Stokes respiration has a characteristic cyclical pattern. Hyperventilation almost always has a psychological basis.

16 Anaemia

1. Introduction

In the fairly recent past, overt paleness of the skin or mucous membranes was said to signify anaemia. This, however, is not always correct. Skin colour, and indeed skin thickness, varies so that the subcutaneous vessels may be more or less prominent. Thus an individual may appear pale without necessarily being anaemic.

When the skin vessels contract (vasoconstriction), as happens in a cold environment or in cases of circulatory insufficiency, the skin becomes paler and at times appears blue with cyanosis; nevertheless the patient may not be anaemic. On the other hand, dilation of the skin vessels or a thin skin may hide an existing anaemia. The colour of the mucosal surfaces provides a somewhat better guide to anaemia than the colour of the skin. The conjunctival membrane of the lower eyelid is a useful guide in this respect (Figures 16.1a and b). In states of anaemia, paleness of the conjunctival membrane is obvious, (though care must be taken to be aware of local misleading patterns, such as inflammation in conjunctivitis). The colour of the mucosal surfaces of the mouth, especially the gums, is also a reasonable indicator of the presence of anaemia.

The characteristic red colour of arterial blood is due to oxygenated haemoglobin (oxyhaemoglobin). Venous blood, being de-oxygenated, has a bluish-purple hue. Practically all the haemoglobin is found in the red blood cells (*erythrocytes*), and anaemia can thus be thought of for most practical purposes as a consequence of a deficiency in erythrocyte numbers or haemoglobin concentration.

Learning outcomes

After studying this chapter, the student should be able to:
- describe the production and function of both the red blood cells and haemoglobin;
- describe two types of anaemia and three main causes of each type;
- outline the clinical picture of a patient with anaemia.

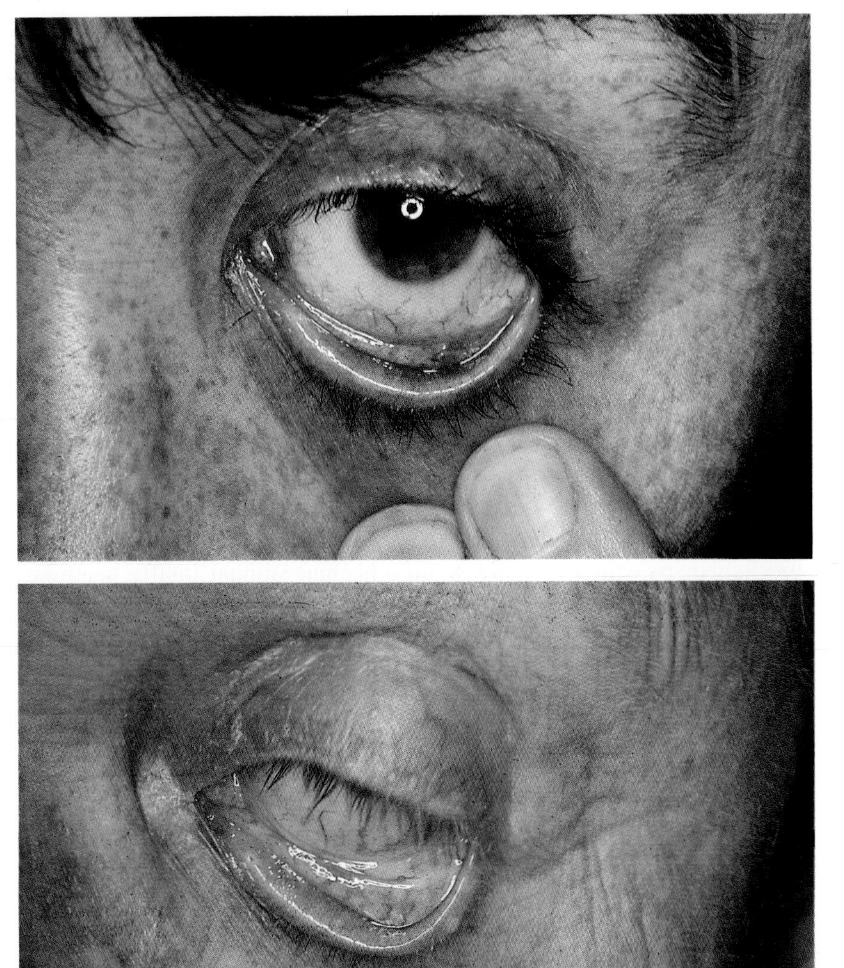

Figure 16.1a
Normal colour of mucous membranes

Figure 16.1b
Anaemia

2. Haemoglobin

Haemoglobin is a relatively large, protein-linked molecule. It contains iron which allows it to bind large volumes of oxygen for transportation to the tissues (this binding capacity is approximately 20 times greater than that of blood plasma). A disorder of this haemoglobin molecule or a deficiency of iron will lead to anaemia. One cause of iron deficieny anaemia is an unbalanced diet lacking in iron-containing foods such as meat, green vegetables etc. Another is chronic blood loss due to, for example, haemorrhoids, gastric ulcers, or menorrhagia. Intestinal malabsorption of iron is a third important cause.

Measurement of haemoglobin, red cell size and shape, total body iron, iron-binding capacity, and bone marrow function can usually classify iron deficiency anaemias.

Red blood cells, or *erythrocytes,* are manufactured in the bone marrow, especially that of short and flat bones. The process is complicated, and many phases are recognised before the erythrocytes eventually lose their nuclei prior to release into the circulation. The average life span of a mature erythrocyte in the circulation is remarkably long, at around 100 days.

Erythrocyte formation (erythropoesis) also requires vitamin B-12, which is found in nutrients such as dairy products, meat, and fish. This is made suitable for reabsorption in the small intestine by the 'intrinsic factor', a protein produced in the gastric glands. The hormone erythropoetin, produced in the kidneys, stimulates the formation of erythrocytes, and anaemia is therefore a not-unexpected consequence of severe renal failure.

An enlarged spleen which becomes overactive in its sequestration of red cells may also bring about anaemia, though the mechanism by which hypersplenism leads to anaemia is not completely understood. The pooling of large volumes of blood from the circulation together with excess red cell destruction is only a partial answer.

Pulmonary dysfunction may actually increase red cell production. In chronic oxygen deficiency due to pulmonary disorders erythropoesis is stimulated, and thus in chronic lung disease (such as chronic bronchitis and emphysema) a state of excess numbers of red cells exists. This is called secondary polycythaemia or erythrocytosis (primary polycythaemia is a neoplastic process where too many red cells are produced). In healthy individuals the ratio of the concentration of haemoglobin to the volume of erythrocytes is fixed, and offers a guide to whether there is a deficiency in haemoglobin or erythrocytes, but a more exact indicator of anaemia is the determination of the Mean Corpuscular Haemoglobin. Examples of the commoner types of anaemia follow.

3. Two types of anaemia

Deficiency in erythrocytes

A deficiency in erythrocytes may be caused by a lack of vitamin B-12. The vitamin may be improperly absorbed due to an intestinal disorder or as a result of liver disease such as hepatic cirrhosis. The main cause, however, is lack of intrinsic factor in the stomach, and this can occur after partial or total gastrectomy, especially if the body itself has produced auto-antibodies against intrinsic factor. This is true *pernicious anaemia* which, at one time a fatal condition, can now be readily treated by injections of vitamin B-12.

In renal disease, the production of erythrocytes is disturbed by the absence of production of erythropoetin. Hypersplenism and anaemia have already been discussed.

In *aplastic anaemia,* bone marrow failure causes failure of the production of red cells (and also of leucocytes and blood platelets). The marrow may become almost totally acellular. Aplastic anaemia may be of unknown cause (idiopathic), or secondary to recognisable causes, including chemical toxicity (sometimes in therapeutic situations when cytotoxics are being administered), radiation, and marrow replacement by neoplasm.

Red cells can be prematurely destroyed by haemolysis (*haemolytic anaemia*). Haemolytic anaemia may arise as a result of some abnormality of red blood cell shape (such as sickle cells, Figure 16.2), by enzyme deficiency, by mechanical damage (heart valve prostheses), and sometimes by excessive exercise, causing erythrocyte trauma.

In clinical situations care must always be taken to avoid haemolysis as a result of incompatible blood transfusion.
Rhesus incompatibility in pregnancy leading to haemolysis in the fetus must be prospectively monitored.

Deficiency in haemoglobin

Haemoglobin deficiency is often caused by iron deficiency due, for instance, to malnutrition, impaired absorption, and/or increased need during pregnancy. It may also be caused by chronic blood loss, as from the gastro-intestinal tract in gastric ulcers, intestinal malignancy or haemorrhoids. Excessive menstrual loss is another common cause.

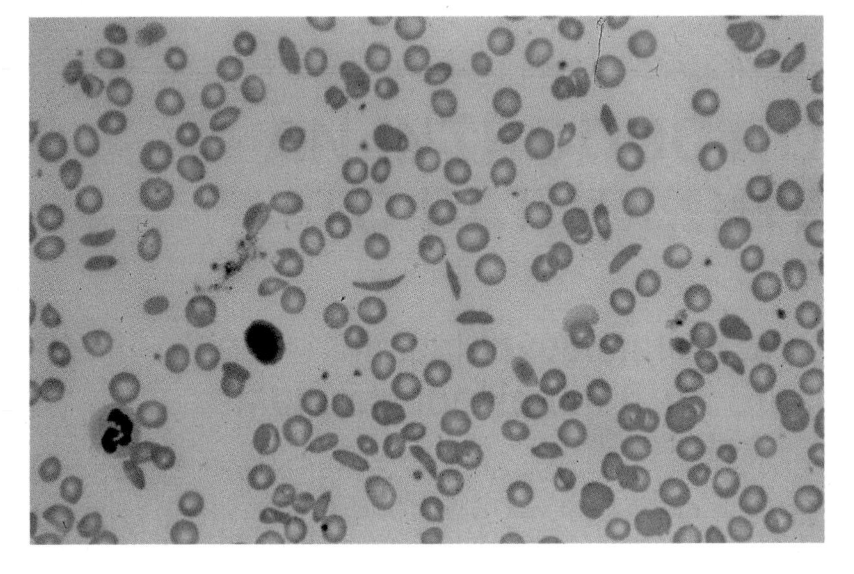

Figure 16.2
Blood film showing sickle cells

4. Signs and symptoms of anaemia

A state of slowly-developing anaemia eventually leads to a chronic oxygen deficiency. The patient complains of fatigue, tires easily, and sometimes also experiences shortness of breath. Other symptoms include pounding of the heart, buzzing in the ears (tinnitus), and - especially alarming to the patient - 'scotomata' (short scintillating lights or fleeting small images in the patient's vision). Clinical signs often include brittle nails, loss of hair and skin fissures (notably at the angles of the mouth). All these signs point to a lesion in epithelial metabolism. Iron deficiency anaemia may lead to difficulty in swallowing (Plummer-Vinsom or Patterson-Kelly-Brown syndrome). Vitamin B-12 deficiency may cause smoothness and pain in the tongue. Peripheral and spinal neuropathy may result in altered peripheral sensation (especially tingling in the legs), and at times to paralysis.

5. Summary

Blood is manufactured in the bone marrow cavities. Erythrocytes contain haemoglobin which effectively binds oxygen. An assessment of anaemia can be gained from examination of mucosal surfaces. The main broad causes of anaemia are:

- erythrocyte deficiency;
- iron deficiency.

17 Cyanosis

1. Introduction

A prime necessity for human survival is the availability of oxygen. Only with the help of oxygen can energy-rich compounds be metabolised to release the energy necessary for normal body function. This chapter addresses situations in which the body lacks oxygen and the ways in which this presents.

Learning outcomes

After studying this chapter, the student should be able to:
- explain the concept of cyanosis;
- describe the occurrence of cyanosis;
- identify the difference between peripheral and central cyanosis and the causes of each type.

2. Causes and indications of cyanosis

When the body is deprived of oxygen, its systems fail. The supply of oxygen to the brain is vital and deprivation for even a few minutes may prove fatal. When atmospheric air enters the lungs, the oxygen passes through the walls of the alveoli into adjacent capillaries and thus into the bloodstream. Haemoglobin in the red cells has the facility to bind oxygen to form oxyhaemoglobin, which has a characteristic bright red colour. This oxygenated haemoglobin is then transported via the blood to the peripheral tissues for utilisation by the tissue cells for normal metabolism. The haemoglobin is then deprived of oxygen, and this gives the blood a bluish colouration. Arterial blood is thus bright red, while venous blood is bluish. (The veins on the back of the hand are a good example of the colour of venous blood). As the capillary blood is well oxygenated, a healthy individual has pink mucosal surfaces in normal circumstances. When, however, the skin and mucosal surfaces have a bluish tinge, *cyanosis* is said to be present. This is usually an indication of an underlying disease process.

For cyanosis to manifest itself there must be a relatively large volume of reduced haemoglobin in the arteries or capillaries. In fact approximately one-third of the total haemoglobin (about 3mmol/litre) must be in the reduced form for cyanosis to occur. In an anaemic patient with too low a level of haemoglobin it is unusual for cyanosis to be present, no matter how poor the oxygen supply may be. Conversely, in a patient with an abnormally high level of haemoglobin, such as occurs in the condition called *polycythaemia*, cyanosis is seen with even minimal exertion. Sometimes it is even seen when the patient is at rest and not suffering any oxygen deficiency.

Therefore, without a knowledge of the patient's level of haemoglobin, it may actually be wrong to conclude from the presence of cyanosis that the patient is suffering from an oxygen deficiency.

Two types of cyanosis are recognised:
- peripheral cyanosis;
- central cyanosis.

The difference between the two types of cyanosis has to be established before the underlying disease process can be diagnosed.

3. Peripheral cyanosis

Peripheral cyanosis, as the name implies, is seen only at the periphery of the body. The term *acrocyanosis* is now less commonly used for the same condition. Peripheral cyanosis is observed at the fingertips, toes, tip of the nose, ear lobes, and lips. It results from a stagnated local peripheral circulation in the capillary areas involved, which results in the local capillaries containing blood low in oxygenated haemoglobin and high in reduced haemoglobin. This need not always be a pathological sign. Physiologically, if there is local vaso-constriction such as in a cold environment the effect is a normal homeostatic mechanism to prevent excess heat loss.

In true circulatory failure, a partitioning of the circulation is a compensatory mechanism. The skin and extremities are temporarily denied normal circulation in order to guarantee an adequate supply to vital centres such as the brain, heart, lungs, liver, and kidneys. These organs thus maintain a near-normal oxygen supply. Examples of such circulatory failure include blood or fluid loss, sepsis, some forms of cardiac failure, and also peritonitis, where blood is sometimes 'pooled' and becomes unavailable to the general circulation because it is trapped in the abdominal cavity.

The flow of blood to the periphery will then be so reduced that peripheral cyanosis, showing circulatory failure, will become manifest.

4. Central cyanosis

In *central cyanosis* all of the circulating blood is de-oxygenated and contains too much reduced haemoglobin. As the lungs are responsible for oxygen transport into the blood, most causes of central cyanosis arise from respiratory disorders. (Certain examples of poisoning which prevent oxygen binding provide other rare causes.)

Disorders of the upper respiratory tract (pharyngeal tumours, tongue tumours and laryngeal tumours) and lower respiratory tract (lung tumours, pulmonary emphysema, chronic bronchitis and bronchial carcinoma) can obstruct the airways sufficiently to cause cyanosis. In pulmonary oedema, oxygen cannot enter the circulation through the alveoli because these are full of fluid, so central cyanosis occurs. A good guide to whether cyanosis is central or peripheral is the presence of a cyanosed tongue, which indicates central cyanosis.

5. More complex types of cyanosis

In some cases central and peripheral cyanosis can co-exist. For instance, in severe bronchopneumonia which results in central cyanosis, the complication of serious sepsis and resulting hypotension brings peripheral cyanosis as well. Massive pulmonary thrombo-embolus provides another good example. Obstruction of the main branches of the pulmonary arteries by thrombo-embolus causes central cyanosis by severely reduced blood supply to the lungs from the heart. The same set of conditions causes circulatory collapse and peripheral cyanosis.
Babies born with congenital heart disease may have a cardiac lesion such that volumes of shunted blood pass from the right side of the heart to the left, missing out the oxygenation cycle in the lungs. As this blood has not absorbed oxygen, central cyanosis develops. Such babies are markedly cyanosed and are commonly referred to as 'blue babies'. This is an example of central cyanosis caused not by a pulmonary disorder but by a cardiac defect.

6. Summary

Cyanosis is a clinical sign which generally indicates oxygen deficiency and excess reduced haemoglobin in the circulation. Cyanosis is broadly divided into two types:

- peripheral cyanosis;
- central cyanosis.

The causes of both types have been illustrated, and the possibility of mixed types has been recognised.

Icterus or jaundice

18

1. Introduction

Icterus or *jaundice* is a yellowish discoloration of the skin and mucous membranes caused by *bilirubin* pigment from the blood (Figure 18.1).

Learning outcomes

After studying this chapter, the student should be able to:
- explain the term jaundice or icterus;
- describe the breakdown of bilirubin;
- identify the different types of jaundice and their causes;
- list and describe the signs and symptoms of a patient with obstructive jaundice and those of a patient with jaundice caused by haemolysis.

2. Causes of jaundice

Bilirubin is a pigment produced by the breakdown of haemoglobin. In the spleen and liver, haemoglobin is released from the old erythrocytes (at about 100 days), after which it is converted into bilirubin. Bilirubin is only slightly soluble in aqueous solution, and it is only because the plasma is rich in proteins (especially colloids) that the bilirubin can be

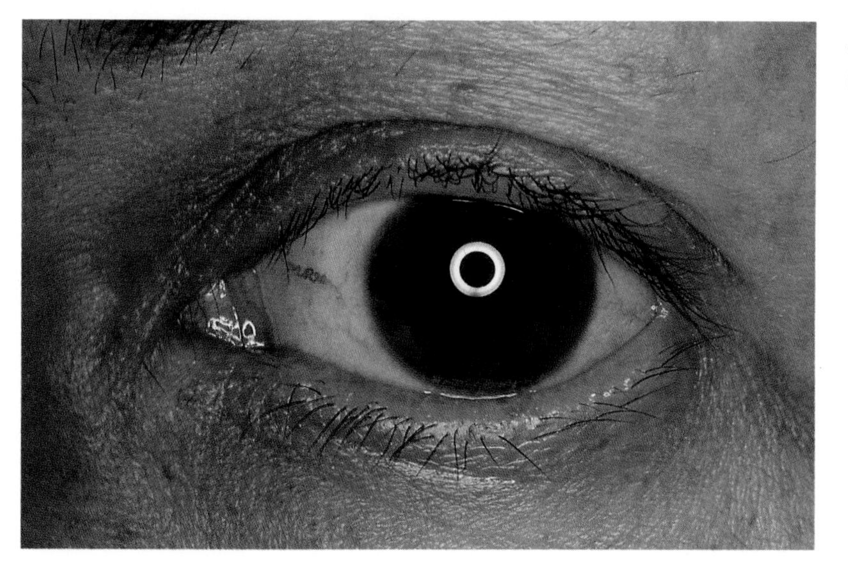

Figure 18.1
Yellowness of the sclera in jaundice

transported in the circulation to the liver. In the liver the bilirubin is made water-soluble by removing attached proteins and adding molecules of glucuronic acid. In this way the originally unconjugated and slightly water soluble bilirubin is converted into conjugated water soluble bilirubin. This can then be excreted as bile into the gall bladder and thence to the small intestines. On passage through the intestines it is further converted to urobilin and stercobilin (Figure 18.2).

Stercobilin is non-soluble and consequently cannot pass through the intestinal wall. It is secreted in the faeces and contributes to their coloration. Urobilin, however, is fully absorbed through the intestinal wall and carried via the portal vein back to the liver, where it is partially re-utilised by being reconverted to bilirubin to complete the cycle. This process constitutes the *entero-hepatic cycle*. The major portion of the urobilin, however, is not metabolised in the liver but enters the blood circulation to be eventually

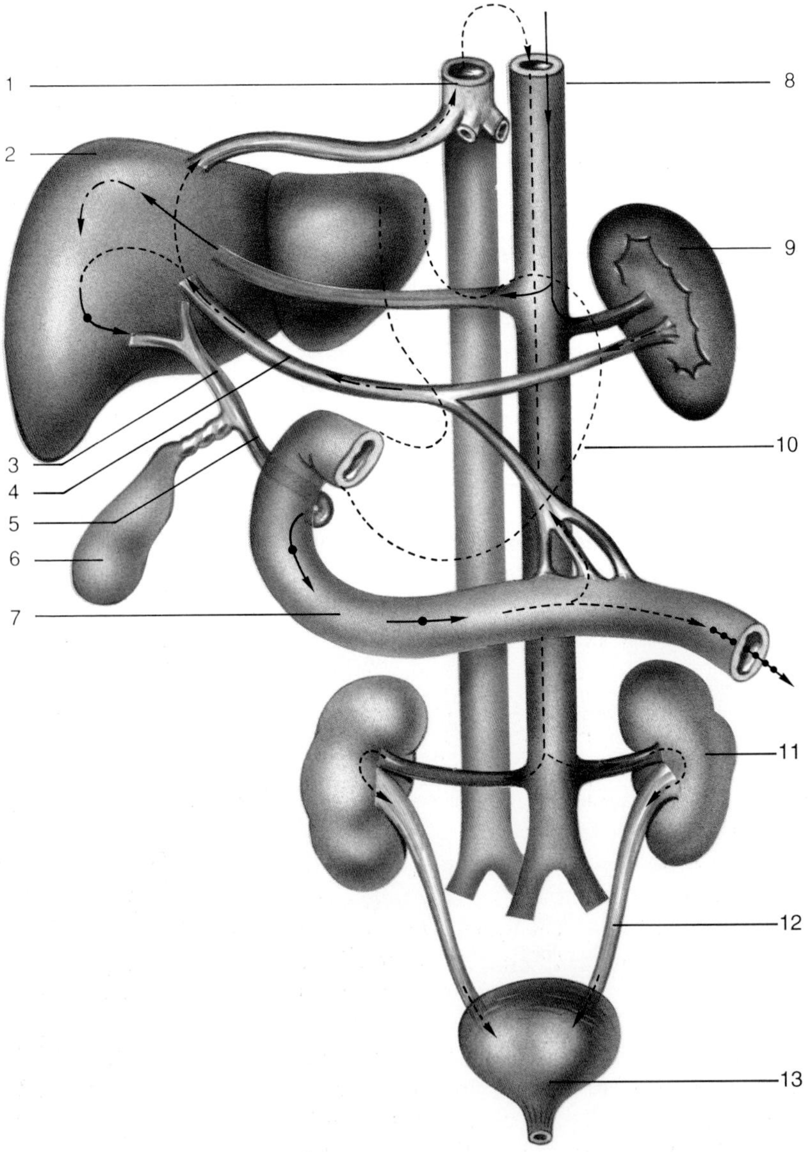

Figure 18.2
Formation, cycle and excretion of bilirubin and urobilin

1. *inferior vena cava*
2. *liver*
3. *common hepatic duct*
4. *portal vein*
5. *common bile duct*
6. *gall bladder*
7. *duodenum*
8. *aorta*
9. *spleen*
10. *outline of stomach*
11. *kidney*
12. *ureter*
13. *urinary bladder*

excreted into the urine. The familiar yellowish colour of urine is partially due to this urobilin.

Although there is always some unconjugated bilirubin in the blood, the amount is normally insufficient to give visible yellow coloration to the skin and mucosal membranes.

Thus, under normal circumstances human beings are not visibly jaundiced, and clearly visible jaundice is an important clinical sign to recognise because it implies a significant underlying disorder.

There are two common mechanisms which bring about jaundice:

1. The rate of production of unconjugated bilirubin is higher than normal.
2. There is excess conjugated bilirubin in the blood and consequently in the skin and mucosal membranes (and in all the other organs too).

3. The two main types of jaundice

a. Increased unconjugated bilirubin in the blood

When the amount of unconjugated bilirubin in the blood increases abnormally, either the supply must be excessive or hepatic metabolism is insufficient.

In the former situation, excessive quantities of haemoglobin derived from destroyed red blood cells are converted into the unconjugated form of bilirubin. This excessive red cell destruction is seen in cases of haemolytic anaemia (see Chapter 15). The result is *haemolytic jaundice*.

The otherwise healthy liver will attempt to receive and excrete as much of this bilirubin as possible. As a result the bile becomes markedly viscous and, due to stasis, gallstones commonly form (cholelithiasis). These gallstones contain a high proportion of bile pigments and are referred to as pigment gallstones. In the intestines the bilirubin is converted into stercobilin and urobilin as usual. The urobilin is carried back in the entero-hepatic cycle to the already overloaded liver which, obviously, can only metabolise a fraction of the urobilin available. A large amount is subsequently excreted in the urine as urobilinuria, and the resulting urine has a characteristic mahogany colour (Figure 18.3).

In hepatocellular disorders such as hepatitis or cirrhosis, the processing of unconjugated bilirubin will be insufficient and large amounts of unconjugated bilirubin therefore remain in the blood, resulting in jaundice.

Despite the impaired function of the liver, some of the unconjugated bilirubin will be metabolised, and the urobilin so produced and excreted into the intestines will be recirculated back to the already malfunctioning liver. The impaired liver becomes unable to handle this additional load so most of this urobilin ends up being excreted in the urine. Urobilinuria is again the outcome and is an important marker in liver disease.

b. Increase in conjugated bilirubin in the blood

This arises when conjugated bilirubin, which must have already passed through the liver cells, enters the blood and thence, due to its water solubility, passes into the urine.

This form of jaundice arises as a result of biliary tract obstruction, and is therefore called *obstructive jaundice*. The site of obstruction may vary and may involve either major or minor bile ducts. Obstruction of the major bile ducts,

Figure 18.3
Urobilinuria - mahogany coloured

caused for instance by gallstones, tumour, abscess, or inflammation, is often almost total. Minor bile duct obstruction due to hepatitis, cholangitis, or the toxic effects of poisons or medication is not so complete. In the first case all the soluble bilirubin re-enters the blood and is excreted in the urine. The resulting urine is very dark (like dark tea) and when shaken froths excessively. As a result of the total obstruction the intestines receive virtually no bilirubin so neither stercobilin or urobilin is produced. As a consequence the faeces lack normal coloration and appear as pale as putty. In obstruction of the minor ducts, especially if there is hepatocellular damage, a more mixed pattern of jaundice appears.

4. Signs and symptoms of jaundice

Examination of the sclerae is probably the most reliable clinical means of establishing the presence of jaundice. Normal sclerae are white; jaundiced sclerae are obviously yellow. Yellow, fatty depositions, especially in the elderly, may confuse the unwary.

Examination of abdominal skin will usually reveal the presence of jaundice, but care must be taken to be aware of local lighting conditions to avoid false interpretation.

In obstructive jaundice excessive amounts of bile salts enter the bloodstream. This results in intractable skin irritation and itching. Bradycardia is also not uncommon.
The froth on the urine is also a result of the presence of bile salts which have the effect of reducing surface tension. Signs and symptoms of the underlying causative conditions must also be sought.

5. Summary

After passage through the hepatocytes in the liver, bilirubin, a waste product of haemoglobin, is water soluble. In cases of obstructive jaundice, this conjugated bilirubin, together with bile salts, accumulates in the tissues and causes jaundice, very dark urine, putty-coloured faeces, and itching. In cases of excessive haemolysis, unconjugated non-soluble bilirubin accumulates and jaundice develops. The urine does not contain bilirubin but does contain urobilin from decomposed bilirubin and has a mahogany hue.

Pain

19

1. Introduction

Among the various symptoms experienced by patients, pain is the most common and most distressing for the individual. In general terms, about 50% of all patients present complaints about some form of pain. Most people rightly relate pain to some underlying disease process. They know that although pain is experienced subjectively, the underlying cause of it is objective.

Learning outcomes

After studying this chapter, the student should be able to:
- explain pain as a symptom;
- describe the path of the pain stimulus;
- identify the different types of pain;
- list the various causes of pain;
- describe the characteristics of pain.

2. The experience of pain

The subjective experience reflects the patient's emotional state of mind, personal circumstances, and even immediate environment.

A sense of safety, security, rest, and understanding may considerably alleviate pain, whereas fear, anxiety, tiredness, insecurity, and discomfort will aggravate it. Recognition of the patient's reaction to pain is therefore important when forming an objective clinical judgement.

The mechanism whereby stimulus of a peripheral pain receptor eventually leads to a sensation of pain is not fully understood. There are indications that in specific pathological processes (such as inflammation), pain is triggered by local release of very small amounts of active substances such as prostaglandins and bradykinins to which neural tissue is extremely sensitive. Whether the activity of these substances acts primarily on nerve endings or directly on nerve fibres is less well understood, but the impulse travels via the dorsal horns of the spinal cord to the hypothalamus. The received stimulus leads to central release of compounds (such as encephalins and endorphins) from the hypothalamus, which effectively excite the brain so that pain is perceived as such. Pain is almost always accompanied by other emotional states, which may include anxiety, anger, despair, or depression. Some individuals in pain become negative and defensive while demanding attention and help, others react with aggression. The individual reaction to pain is very varied and interpretation takes significant clinical skill.

The site of perceived pain may be clinically misleading.
Often the location at which pain is experienced is different from the site of the causative stimulus. This is known as *referred pain* and usually relates to patterns of innervation in the developing fetus. A further complicating factor is that not all body tissues are sensitive to pain.The interiors of the liver, spleen, heart, lungs and brain do not possess pain receptors, but their overlying membranes (peritoneum, pericardium, pleurae and meninges) are pain sensitive. Bones lack pain receptors but the periosteum is pain sensitive. The skin, subcutaneous tissues, muscles and ligaments are richly supplied with pain receptors and are very sensitive to pain - a useful protective mechanism. Finally, duration of pain markedly alters its perception. Disease processes are such that pain originating in a serosal mechanism (like the peritoneum) or in a muscle or tendon will probably be constant and continuous. A forceful contraction of a hollow organ (such as a bile duct attempting to overcome an obstruction caused by a stone) will cause rapid onset of severe, but relatively short lasting, colicky pain. Pain originating from a contracting uterus in labour will have a more undulating character.

3. Classification

The large number of disorders in which pain is a significant factor can be classified by a combination of anatomical site and type of neural irritation:
- inflammation in general;
- spasm in general;
- oxygen deficiency in general;
- irritation of a serosal membrane;
- irritation of the skin, muscles, joint capsules, and ligaments.

a. Inflammation

Virtually all inflammatory conditions are accompanied by pain of varying severity. As already stated, local release of substances, such as prostaglandins and bradykinins, which trigger pain play a role in the process.These locally released substances also lead to local vasodilation, vessel permeability, and oedema, and thus contribute to producing the five classical characteristics of inflammation:
- rubor (redness);
- calor (heat);
- dolor (pain);
- tumor (swelling);
- functio laesa (impaired function).

b. Spasm

A spasm causes colicky, spasmodic pain and it has an unpredictable rhythm of recurrence. Such

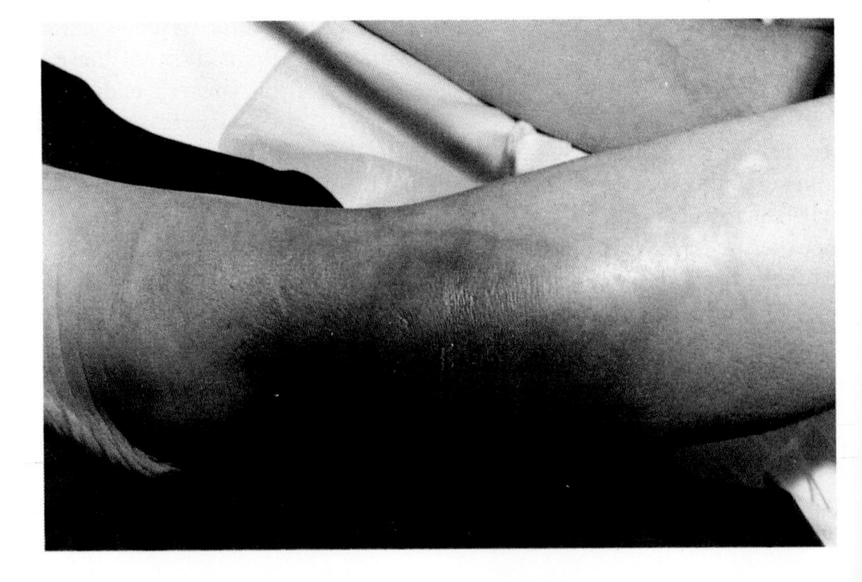

Figure 19.1
Acute inflammation with clearly recognisable symptoms

colicky pain is common in obstruction of a hollow viscus, such as a biliary duct occluded by a gallstone or an obstruction of the ureter by a urinary calculus. The pain is severe and the patient, in an attempt to find a position of relief, demonstrates marked restlessness.This motor restlessness is an important clinical sign to recognise.

c. Oxygen deficiency causing ischaemia

Insufficient blood supply leading to tissue oxygen deficiency brings on a sharp, continuous pain.Typical is the tight, strangling nature of the pain found in angina pectoris and myocardial infarction, which is caused by ischaemia of the myocardium due to insufficient oxygen supply from the narrowed and diseased coronary arteries. The pain experienced in intermittent claudication is similar, forcing the sufferer to cease activity until sufficient oxygen reaches the legs by the reduced blood flow through the narrowed leg arteries. A comparable situation involving the intestines may occur as a result of narrowing of the mesenteric arteries.

d. Irritation of serous membranes

Assessment of pain experienced in the serosal membranes forms a significant part of clinical examination. One notable feature is the increased severity of pain when the membrane is moved.

Pleural pain, for example, is aggravated by deep breathing or coughing, which the patient thus tries to avoid.In peritonitis the patient avoids movement of the peritoneum. In examination this is known as 'guarding', and sudden removal of the examining hand from the abdomen causes 'release pain', a further typical sign of peritoneal irritation. Peritonitis has many causes, including perforated gastric ulcer and acute appendicitis.
In meningitis, the existing pain is made worse by stretching the meninges when the head is flexed on the neck and the patient automatically resists this, demonstrating nuchal rigidity. Similar stretching movements aggravate existing pain in lesions involving the periosteum, muscles, and joint capsules. Knowledge of the detailed anatomy and nerve supply provides specific diagnostic information.

4. Summary

Pain is a subjective reaction to an objective stimulus and therefore varies between individuals.This chapter has outlined the pathways of pain transmission in the nervous system.The general causes of inflammation, spasm and ischaemia have been outlined, and more specific anatomical sites of pain (pleura, peritoneum, meninges, joint capsules, muscle, and skin) have been introduced. The importance of detailed clinical history and examination is stressed.

20 Pain in the chest

1. Introduction

Most people suffering from chest pain, whether of long-standing or of recent origin, are anxious that there may be an underlying heart disorder. Luckily, in most cases this is a groundless fear and the cause of the pain lies elsewhere. Chest pain may indicate disorders arising not only in the thorax but also from below the diaphragm or in the head and neck (the fact that pain may be felt in sites other than location of the root cause has already been noted). Chest pain therefore varies widely in its perception and may have many causes.

Learning outcomes

After studying this chapter, the student should be able to:
- identify the structures from which pain in the chest can be triggered;
- describe the characteristics of chest pain;
- explain the differences in perceived pain between some important disorders.

2. Thoracic wall

Many examples of chest pain originate in the chest wall muscles, nerves, and skin. Any irritation of intercostal nerves will lead to local pain (*intercostal neuralgia* is the general term).The inflammation and vescicles along the site of nerves, caused by herpes zoster virus in shingles, is a good example (Figure 20.1). Chest pain is accentuated by movement, including deep breathing. As the virus attacks and manifests along a single spinal nerve distribution, the affected area can be precisely located even before the characteristic rash appears. Local trauma to a thoracic vertebra is another possible cause of intercostal pain.

3. Ribs

A bruised rib is locally very painful. If fractured, crepitus between the ends of the broken bone may be felt, or heard using a stethoscope.

A pathological fracture of the ribs may result from a metastatic deposit from a distant malignant neoplasm. Common sites of origin include lung, breast, kidney, and prostate. Local tumours may infiltrate intercostal nerves and cause pain. Less serious in outcome but worrisome in presentation to the patient are lesions such as local *costochondritis (Tietze's syndrome)*, the cause of which is not fully understood.

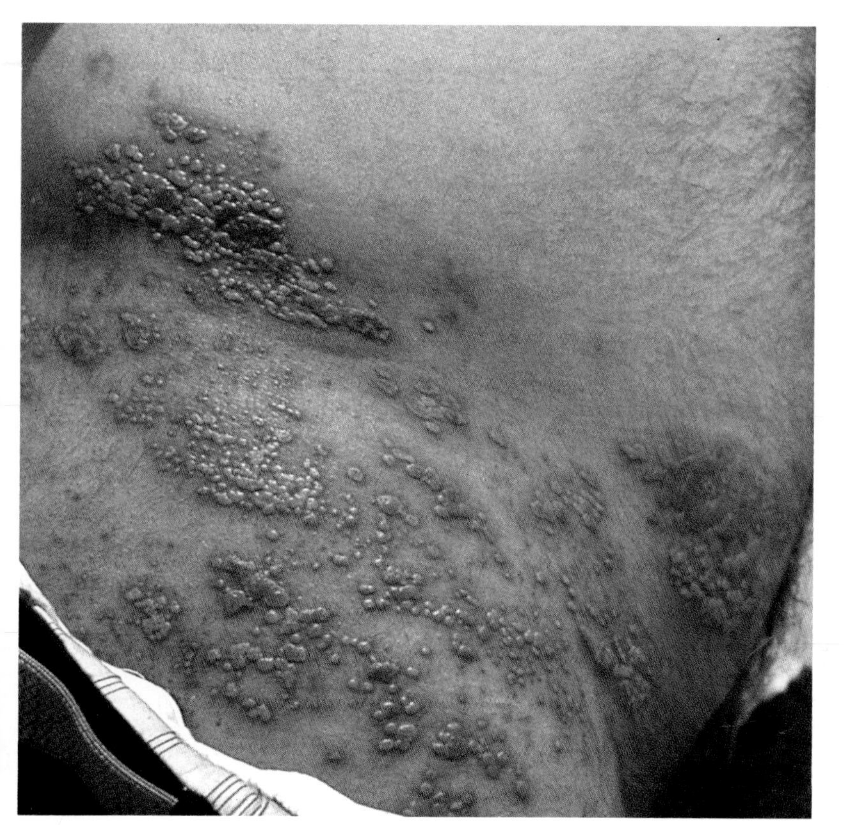

Figure 20.1
Herpes zoster (shingles)

4. Vertebrae

The spinal nerve routes coming from the spinal cord leave the vertebral column between two vertebrae then progressively divide and subdivide peripherally. Irritation of sensory nerves at vertebral level can have many causes, including congenital malformation of the vertebrae, intervertebral disc degeneration, inflammatory and infective conditions, and local infiltration by malignant disease (Figure 20.2). Appropriate signs and specific symptoms together with additional investigations will be required to arrive at a specific diagnosis.

5. Lungs and pleurae

The substance of the lungs itself is not sensitive to pain. Pneumonia, bronchitis, or bronchial carcinoma do not cause pain unless the overlying pleura is involved in the pathological process. An illustrative example is *pulmonary embolus* where the pleura overlying the segment of lung deprived of blood circulation is subject to ischaemic damage and subsequent inflammation, and thus pain occurs.With every breath, there is movement between the two inflamed layers of the pleura. As a result, the patient avoids deep breathing.The pleural rub can be readily heard using a stethoscope. Later in the disease process a reactive pleural effusion forms and separates the layers of pleura and the pain subsequently disappears. A similar pattern of pleural irritation is sometimes found in pneumonia and lung cancer.

Pneumothorax (Figure 20.3) causes severe, sharp chest pain of sudden origin when air enters the pleural space normally only occupied by a small amount of lubricating fluid. The air may enter from the outside environment as a result of a penetrating chest injury such as a stab wound, or from the lungs when a breach occurs on the visceral pleura, either spontaneously or as a

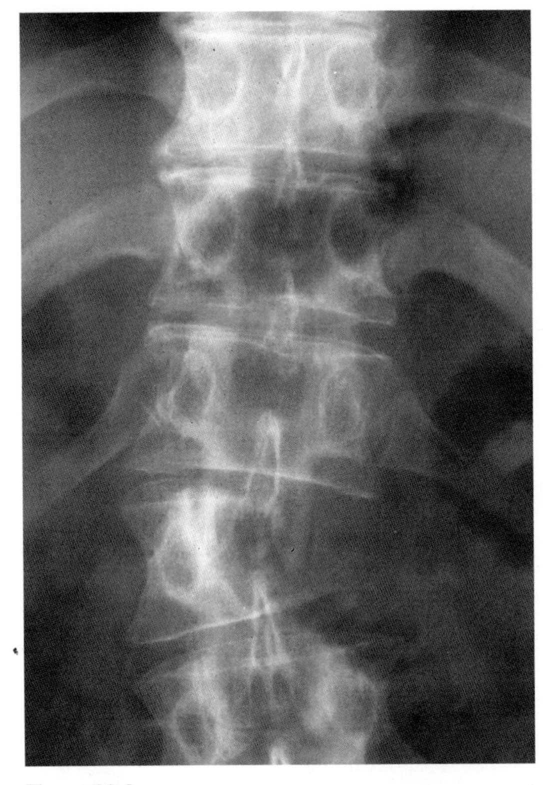

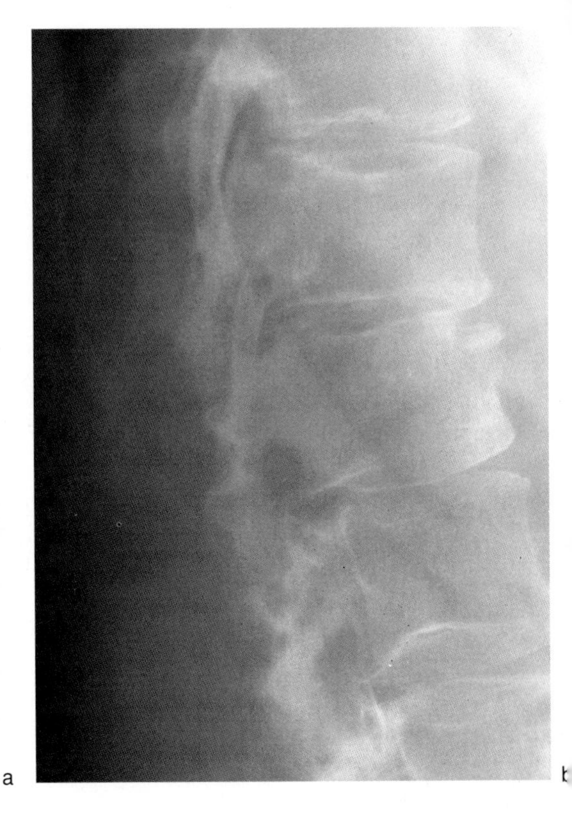

Figure 20.2
Fracture of a vertebra as a result of metastatic neoplasm

a. anterior-posterior view
b. lateral view

result of underlying pathology. The patient experiences sudden onset of pain (often after slight exertion in the spontaneous cases).The pain is unilateral and the ensuing breathlessness is progressive. The normal and relatively negative intra-pleural pressure is replaced by a positive air pressure when the underlying lung collapses and loses some of its respiratory function. On examination the normally slight intercostal depression disappears unilaterally and the chest wall movement lags behind respiration on the affected side. A more serious situation arises when a valve-like mechanism arises at the site of the lesion and more and more air and pressure develop. This constitutes a tension pneumothorax.The increasing volume of air and pressure push the heart, major vessels, and other lung aside and a shift of the trachea is sometimes clinically detectable. A tension pneumothorax is an emergency situation. A large bore needle inserted into the pressurised pleural space will equalise pressure with the external environmental air pressure.

6. Disorders below the diaphragm

The patient with pain in the chest does not generally relate it to an abdominal disorder, but the health care professional must always consider such causes if a disaster is to be avoided.

Cholecystitis (inflammation of the gall bladder) may cause pain at either side of the thorax (Cope's symptom).

Perforation of the stomach or other organ or acute pancreatitis can cause thoracic pain (often on the left).

Appendicitis and peritonitis from other causes (including pelvic, for example salpingitis) can cause diaphragmatic irritation and referred shoulder pain. Even a left colon overdistended with flatus can cause left-sided thoracic pain. In

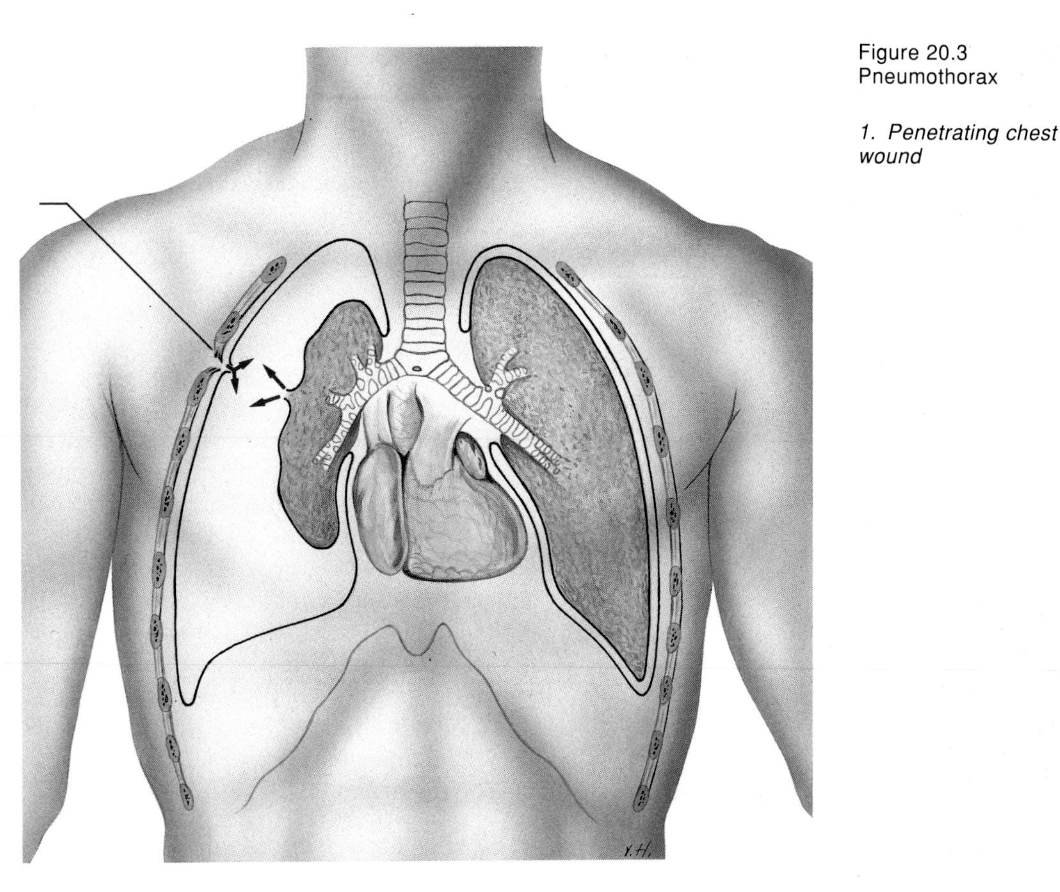

Figure 20.3
Pneumothorax

1. Penetrating chest wound

all these cases and other abdominal conditions specific localising signs and symptoms will aid diagnosis.

7. Oesophagus

The cardio-oesophageal junction functions as a physiological rather than a substantial anatomical sphincter. It prevents gastric contents from regurgitating into the oesophagus. Sometimes a portion of the stomach will herniate into the thorax through the diaphragm, forming a hiatus hernia (Figure 20.4). This can be accentuated by coughing, sneezing, straining, lifting or lying in bed, and it causes gastric content reflux. Pain may commence in the upper abdomen and radiates to the thorax, often the left shoulder.The patient complains of heartburn, belching, and nausea, often alleviated by moving to a standing position. Differentiation from angina pectoris can be very difficult.

8. Heart

Angina pectoris arises in most cases as a result of relative ischaemia in the myocardium due to coronary artery atheroma. Commonly, the left anterior descending branch of the left coronary artery is the site of severe lesions.
The condition is brought about when myocardial demand for oxygen exceeds available supply. This may have an emotional basis such as anger, anxiety, or fright. Physical exertion obviously increases the myocardial need for oxygen. Blood supply to the heart will be relatively reduced if demands elsewhere in the circulation arise, as may occur when moving from a warm to a cold environment, or after a heavy meal.

In angina pectoris the pain has a classical pattern. There is central chest tightness, often described as akin to a heavy oppressive weight or tight band. The pain radiates to the neck,

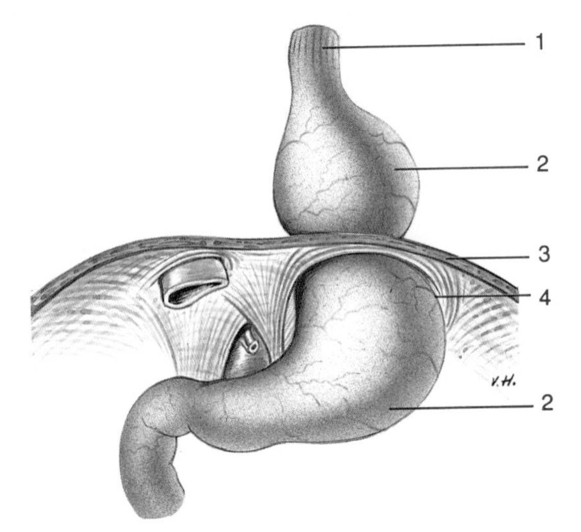

Figure 20.4
Sliding hiatus hernia
Portion of the stomach in the thorax

1. oesophagus
2. stomach
3. diaphragm
4. hiatus in the diaphragm

jaws, and left arm, and there is often also tingling in the left arm. The patient is often anxious, not surprisingly. Rest from activity, and coronary artery vasodilation with substances such as glyceryl trinitrate, usually bring relief in a few minutes. The demand on the myocardium decreases, the oxygen supply increases and with the disappearance of the myocardial ischaemia the pain subsides.

Myocardial infarction occurs when the ischaemia is not relieved and frank myocardial tissue death takes place. This may result from total occlusion of a major coronary artery by thrombus overlying an atheromatous plaque, or it may be a direct result of occlusive atheroma. In such events the patient experiences continuing and more severe pain, and feelings of nausea are common. The blood pressure falls and various cardiac dysrhythmias are common. The damage to the myocardium is permanent and loss of pump ability has occurred. Diagnosis can be made from history, examination, electrocardiography, and cardiac enzyme assay in the blood. Disorders of the pericardium may present as cases of chest pain (pericarditis is one such example).
Aneurysm of the thoracic aorta is a further important cause of chest pain. In contrast to angina pectoris, these last two disorders do not show any alleviation of the pain with rest.

9. Summary

Chest pain is alarming to the patient, who often suspects a heart disorder. However, chest pain more commonly has other causes. These include local disorders such as intercostal neuralgia, rib injuries and vertebral disease. Pulmonary pathology leading to pleural involvement, such as pulmonary embolus or pneumonia, is a further important cause. Disorders in the abdomen (such as hiatus hernia, cholecystitis or peritonitis) may cause pain in the chest. Angina pectoris and myocardial infarction have a classical clinical history. In all cases a careful study of the clinical history and a thorough clinical examination are essential in arriving at the correct diagnosis.

Abdominal pain

21

1. Introduction

The lay patient often complains of 'stomach ache' as a general manifestation of abdominal pain. The presentation of abdominal pain is very varied and the types of underlying disorders are numerous. Like chest pain, abdominal pain may relate to disorders beyond the abdomen. Examples commonly encountered include myocardial infarction, pleurisy, and hiatus hernia. Myalgia or neuralgia originating from neurological disorders at spinal level and affecting the abdominal walls may also present as abdominal pain, as may some general metabolic disorders including dehydration, uraemia, acidosis and other toxic states.

There are many ways to classify abdominal pain. For simplicity this text will consider *acute* and *chronic abdominal pain,* and also those types of abdominal pain which involve colic and those which do not.

Learning outcomes

After studying this chapter, the student should have sufficient knowledge and understanding of:
- the pathological processes of a gallstone (or biliary) colic, and a kidney (or renal) colic;
- the pathological processes of intestinal ileus;
- the clinical features of specific abdominal conditions.

2. Colicky abdominal pain and mechanical obstruction

Colic can be described as acute, paroxysmal, severe pain having a characteristic crescendo/decrescendo pattern. This is usually accompanied by general motor unrest. Such colicky abdominal pain is almost always caused by 'blockage' or stenosis of a hollow viscus, especially of the biliary or urinary tract or the intestines.

Biliary colic is the result of blockage of a biliary duct, by a gallstone or, more rarely, by acute inflammation or neoplasm.The pain, initially experienced in the right upper quadrant of the abdomen, frequently radiates to the right flank and the right shoulder blade (Figure 21.1). Nausea and vomiting are common and may be triggered by food, especially of a fatty nature. The vomit may contain biliary material and the urine often darkens as a result of the presence of bilirubin. Obstructive jaundice may develop.

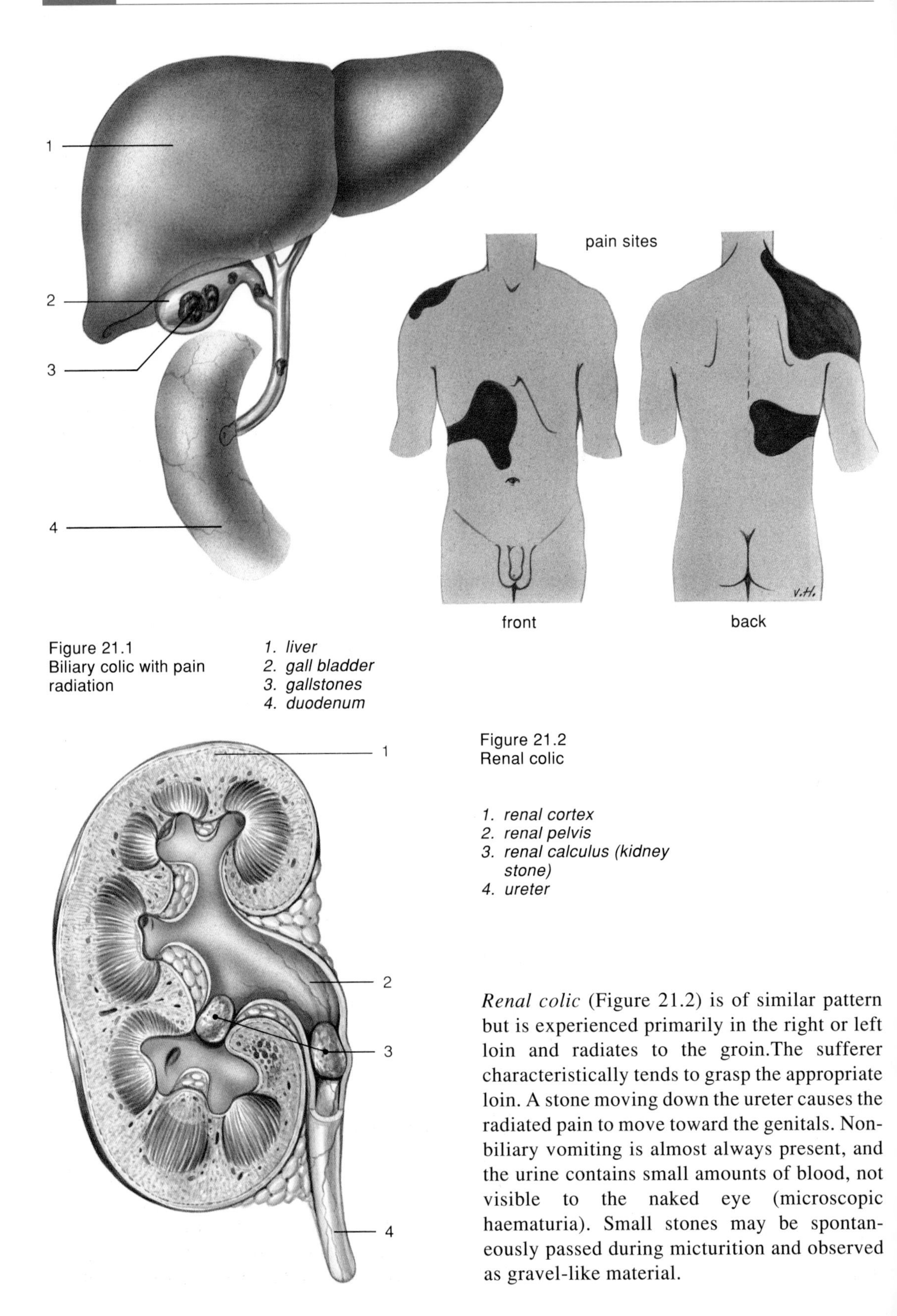

Figure 21.1
Biliary colic with pain radiation

1. *liver*
2. *gall bladder*
3. *gallstones*
4. *duodenum*

Figure 21.2
Renal colic

1. *renal cortex*
2. *renal pelvis*
3. *renal calculus (kidney stone)*
4. *ureter*

Renal colic (Figure 21.2) is of similar pattern but is experienced primarily in the right or left loin and radiates to the groin.The sufferer characteristically tends to grasp the appropriate loin. A stone moving down the ureter causes the radiated pain to move toward the genitals. Non-biliary vomiting is almost always present, and the urine contains small amounts of blood, not visible to the naked eye (microscopic haematuria). Small stones may be spontaneously passed during micturition and observed as gravel-like material.

Intestinal colic is also accompanied by acute pain.The crescendo/decrescendo pattern of increasing pain, alternating with pain-free intervals, is very apparent. Female patients relate intestinal colic to the pain of uterine contractions in labour.The acute pain in intestinal colic is accompanied by an increased peristalsis as the intestine attempts to overcome the underlying intestinal obstruction.As a result of the obstruction only small amounts of faecal matter are passed, and there is little flatus.
The increased peristalsis is visible on abdominal examination and is detectable on auscultation with a stethoscope. Radiographic survey of the abdomen will show fluid levels and indicate whether the intestinal obstruction is small or large (Figure 21.3). Eventually the abdomen becomes distended.The obstruction, if in the small intestine, will lead to vomiting of altered, partially digested food material.In large bowel obstruction the abdominal distension may be more prominent in the lateral aspects and the accompanying vomitus has a faecal odour, but recognisable faecal-like material in the vomitus is rather unusual.

Figure 21.3
Radiograph of the abdomen showing fluid levels

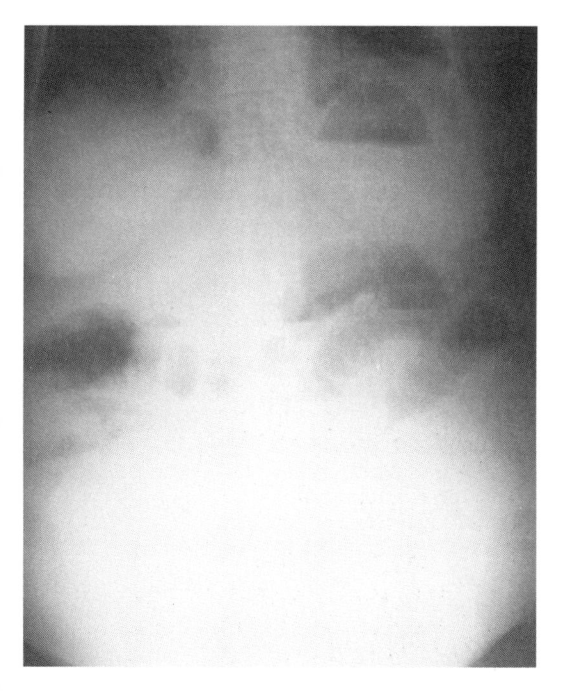

There are many causes of mechanical obstruction. In children, volvulus (twisting) of the intestines and intussusception (invagination of one portion of intestine into another) can cause mechanical obstruction. In all age groups scarring after an inflammatory process may cause occlusion. Similarly, adhesions after previous surgery may result in obstructive bands. Hernias, where loops of bowel become trapped in areas such as the inguinal canal (inguinal hernia) or femoral canal (femoral hernia) can also obstruct and cause mechanical blockage.

A large gallstone in the small intestine can lead to gallstone ileus. Tumours, both benign and malignant, often present as intestinal obstruction. Inflammatory masses, such as occur in diverticular disease, are common causes of large bowel obstruction. Surgical intervention is often required to diagnose and treat intestinal obstruction.

3. Acute abdominal pain, peritonitis, and paralytic ileus

Inflammation and irritation of the peritoneum (*peritonitis*) is clinically striking. The patient is in shock, with a pale appearance. The extremities are cold and the face clammy. Any movement accentuates the pain so the patient lies very still, often protecting the abdomen with the hands.

A patient with peritonitis shows much reduced (or sometimes no) abdominal respiration. Palpation of the abdomen is resisted and the protective response of muscle 'guarding' is observed. In established peritonitis, peristalsis ceases and no bowel sounds are heard on auscultation of the abdomen. As there is no effective peristalsis, no faeces are produced and no flatus passed. This situation is described as 'paralytic ileus'.

The causes of peritonitis are many. Generally speaking, the peritoneum regards most substances as foreign and reacts as if they are irritants. It therefore reacts adversely to bacterial infection, blood, gastro-intestinal juices, bile and many other substances.Thus a ruptured intestine, no matter whether it is due to trauma, infection or tumour, will lead to peritonitis. Similarly, any release of pancreatic juices (in pancreatitis) or blood (from a ruptured liver or spleen or from a ruptured ectopic tubal pregnancy) will lead to peritoneal irritation.

A careful study of the case history, clinical examination and other appropriate diagnostic investigations will be required to make a specific diagnosis. General observations of pulse, blood pressure, respiration, and temperature are essential, as is maintenance of accurate fluid balance. General supportive therapy is paramount until specific therapies, medical or surgical depending on the diagnosis, are implemented. In peritonitis, surgical intervention is often required if the patient is to survive. Lameris (1940) is quoted as stating poignantly, 'The sun must neither rise nor set over an acute abdomen'. Time is always of the essence when dealing with cases of peritonitis.

The diagnosis of acute abdominal pain can tax diagnostic skills. Finding the exact source of a chronic abdominal pain can be even more of a problem. Consideration of the abdomen as *four quadrants* provides some help. Diseases of the liver and biliary tract (and sometimes of the right kidney) are perceived in the upper right quadrant. Disorders of the stomach, spleen, pancreas and left kidney are most likely to be perceived in the upper left quadrant. Stomach conditions often present with umbilical pain, as does appendicitis in its early stages. Renal conditions often show a loin to groin distribution. Disorders of the appendix, right ovary, right ureter, and right fallopian tube often present with lower right quadrant symptoms. Symptoms referred to the lower left quadrant may originate in the left ovary, left ureter, left fallopian tube, or descending colon.

Disorders such as 'irritable bowel syndrome' may present with some, all or none of the above, but they often have some overlying psychosomatic component. Vascular insufficiency, comparable to angina pectoris but caused by atheroma of the mesenteric arteries, gives rise to so-called 'abdominal angina'.

Carcinoma of the stomach may give clinical signs and symptoms indistinguishable from chronic gastric peptic ulcer. Similarly, chronic pancreatitis may mimic pancreatic carcinoma, as may carcinoma of the colon and diverticular disease of the large bowel. Further and often exhaustive, detailed investigations may be unavoidable in establishing the exact underlying pathology of an abdominal pain.

4. Summary

Abdominal pain may present acutely or it may be of a chronic nature. Colic relating to obstruction of hollow organs such as bile ducts, ureters, and intestines has been described. The nature of peritonitis and a few of its commoner causes have been considered, as has paralytic ileus. The problems of diagnosis of chronic abdominal pain have been briefly mentioned and the helpful anatomical concept of abdominal quadrants has been introduced.

Headache

22

1. Introduction

Headache is the commonest complaint expressed by patients to their medical advisors. It is estimated that an average general practitioner is confronted with a new patient complaining of headache every day. As a symptom, headache is a poor guide to the underlying causative pathology. This is partially explained by the fact that most causes of headache originate from areas of the body outwith the skull. Brain tissue itself is not sensitive to pain.The coverings (pia mater, arachnoid mater, and dura mater) are variably pain sensitive.The arachnoid is only slightly sensitive, the pia and dura mater more so (Figures 22.1a and b). Branches of several of the cranial nerves, including the trigeminal nerve, however, are pain sensitive.

Learning outcomes

After studying this chapter, the student should be able to:

- identify anatomical sites which generate headache;
- understand the types of disorder which lead to headache, and give examples of these.

2. Inflammation and infection

High fever is almost always accompanied by headache.This is especially true for children but not unusual in adults. A vague headache is often present in viral infection, and sometimes it is accompanied by vomiting. Local inflammatory conditions in the head presenting with headache include *otitis media* (middle ear infection), in which the sufferer, usually a child, is in much distress and typically grasps the offending ear. Complications include mastoiditis, meningitis, and cerebral abscess, all of which are life-threatening conditions.

Another common cause of headache due to local inflammation is *sinusitis*. This is a common complication of an upper respiratory infection. Frontal sinusitis causes pain just above the eyes. Ethmoidal sinusitis causes pain at the corner of the eyes. In maxillary sinusitis the area of pain is in the face between the upper jaw and the eye. Coughing, straining, and bending tends to accentuate the pain. Patients with sinusitis have a history of upper respiratory tract infection and may have a purulent nasal discharge or a 'blocked' nose.

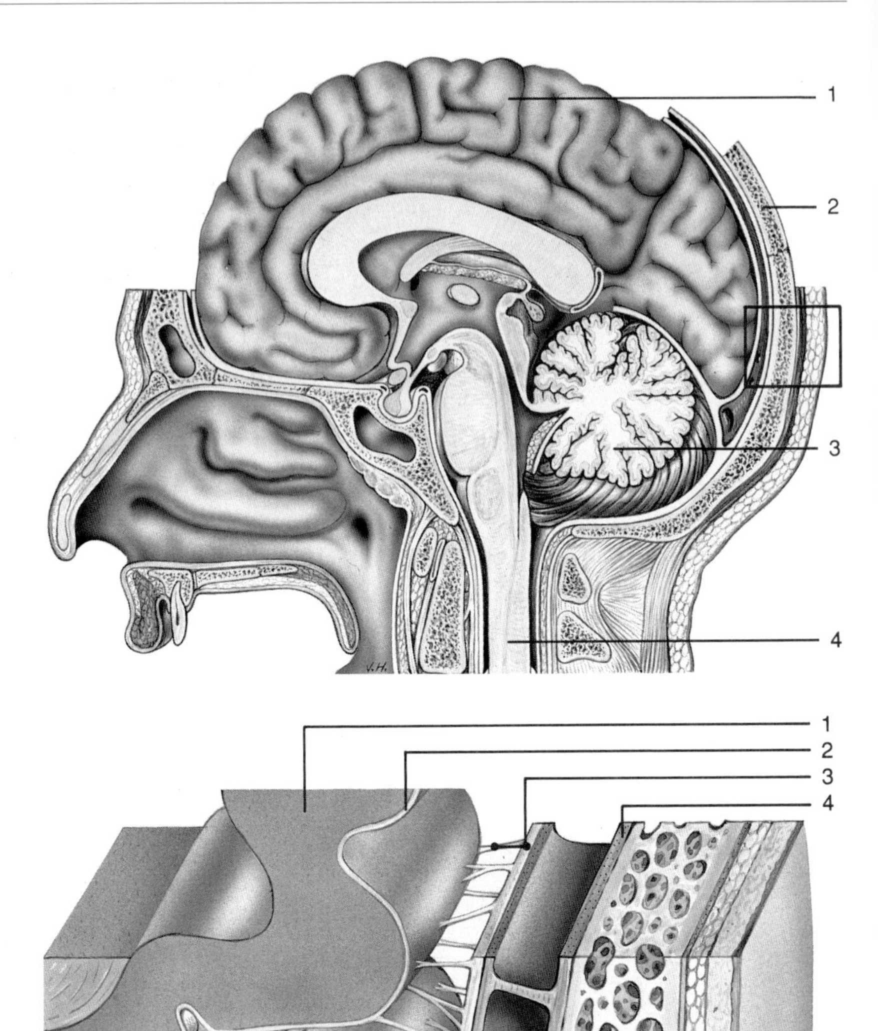

Figure 22.1a
Meninges

1. *cortex*
2. *skull*
3. *cerebellum*
4. *medulla oblongata*

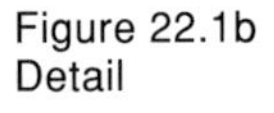

Figure 22.1b
Detail

1. *cortex*
2. *pia mater*
3. *arachnoid*
4. *dura mater*
5. *skull*

Inflammatory conditions of the *eye* are often accompanied by headache. Iritis or iridocyclitis commonly produce headache, unilateral in position if only one eye is involved.

Inflammation of the *temporal artery*, especially in the elderly, causes severe throbbing headache. In this condition a swift and accurate diagnosis is essential as blindness is a frequent complication. A typical history plus tender nodularity of the offending artery, together with a markedly elevated erythrocyte sedimentation rate, are diagnostic indicators, and a biopsy specimen will show giant cell and granulomatous arteritis. Steroid therapy, if administered in time, prevents blindness.

Inflammation of the *meninges* (meningitis) may also cause severe headache and is often accompanied by neck stiffness (nuchal rigidity).

3. Exogenous toxic causes

Domestic central heating has reduced the numbers of cases of chronic carbon monoxide poisoning, but faulty gas flues and poorly ventilated paraffin heaters are still a significant problem. Chronic sub-lethal poisoning by carbon monoxide is an important cause of otherwise unexplained chronic morning headache. The socially deprived, elderly and caravan dwellers are particularly at risk. Car exhaust fumes in a confined space are another well recognised cause of headaches. Suicide attempts by this method are increasing. As the carbon monoxide binds firmly to the haemoglobin to form carboxyhaemoglobin, diagnostic assay of this compound is possible. Once a significant amount of carboxy-haemoglobin is present in the blood the patient exhibits a typical 'cherry-pink' colour. Levels of more than 45% carboxyhaemoglobin are usually fatal.

Many other substances cause headache through excessive exposure, including formaldehyde, lead compounds, and smoke. A 'hangover' as a result of excessive alcohol consumption is a fairly common toxic state. The toxins in this case are the trace terpenes and fusel oils in the beverage which a few hours previously were being praised for their aromatic bouquet.

Sundry medications cause headache as a side effect. Patients with angina pectoris taking vasodilators containing glyceryl trinitrate often experience headache due to extra-cranial vasodilation.

4. Metabolic disorders

Any general metabolic disturbance may be accompanied by headache, caused as a direct consequence of the upset to the normally well-regulated homeostasis. Thus, headache is common in conditions as diverse as diabetes mellitus, liver failure, renal failure, enteritis (often due to the circulation of specific enterotoxins), shock and respiratory failure.

In this category can be included *migraine* although the specific pathogenesis of this condition is still not completely understood. It is certainly recognised, however, that a sudden release of vaso-active serotonin takes place and that an altered response of the extracerebral cranial vessels occurs resulting in a classical migrainous attack. Immediately prior to the headache, the patient feels unexpectedly tired and has visual and, at times, auditory disturbances. The sufferer often also experiences photophobia (hypersensitivity to light). The headache is usually unilateral and at its most severe the victim may vomit. Some females note a premenstrual predilection for migrainous attacks. Several other varieties of migraine like 'cluster-headaches' are recognised.

Where a patient complains of headache accompanied by perspiration, trembling, tachycardia, hunger sensation and aggressive or altered mental state, great care must be taken not to miss hypoglycaemia in a diabetic with insulin overdose.

5. Vascular disorders

As noted in Chapter 19, oxygen deficiency causes pain. Headache following cerebral thrombosis, haemorrhage, or embolus may result from swelling of the ischaemic cerebral structures irritating the meninges. Angiospasm leading to relative oxygen deficiency is another mechanism which causes headache. The changes which take place during migraine attacks, described above, obviously have a significant resultant vascular component.

Headache in anaemia and carbon monoxide poisoning is also partly explained by relative oxygen deficiency.

6. Processes causing raised intra-cranial pressure

The skull is a rigid container of fixed volume. The brain is largely composed of water and is suspended in cerebro-spinal fluid. There is little scope for any intra-cranial expansion without detriment.

When the intra-cranial volume tries to increase as a result of haemorrhage, inflammation, oedema, or tumour (whether benign or malignant, primary or metastatic), the meninges are stretched and irritated. Headache results, often accompanied by vomiting and relative bradycardia (probably as a result of vagal stimulation). Meningeal pain commences as a dull ache which then becomes severe, persistent, and almost unbearable. In this state of *meningismus* (meningeal irritation) the patient lies with head flexed and knees pulled up in an attempt to reduce the pressure. The pain in meningitis and sub-arachnoid haemorrhage is especially severe.

7. Irritation of the cranial nerves

Irritation of a sensory division of a cranial nerve causes headache which is often anatomically localised and therefore provides useful diagnostic information. *Trigeminal neuralgia* is one such example, as are some cases of toothache.

8. Summary

In many cases, headache results from a general systemic metabolic or exogenous toxic condition rather than from a local cranial lesion. Local intra-cranial lesions, including tumour, haemorrhage and inflammation, result in meningeal stretching, irritation and headache. Extra-cranial vascular disorders commonly cause headache. Migraine is an important common condition. Many headaches have a significant psychosomatic component.

Fatigue

23

1. Introduction

A significant proportion of the population regularly complains of fatigue. In the hustle and bustle and occupational stress of contemporary life, fatigue is a very common complaint. It is also a condition which is difficult to define with precision. The term fatigue usually refers to a generalised feeling of tiredness or exhaustion often leading to a lack of energy and decreased efficiency. It is all too easy to overlook those people for whom fatigue signifies an underlying physical or psychological problem. But it is essential to try to detect the presence of any such condition, and to avoid cursory dismissal of the patient.

One of the most difficult problems for any health care professional is the assessment of this symptom and the decision of whether or not it has real significance.
Fatigue often accompanies a physical disorder; however, it is also common as a symptom of an underlying emotional or psychiatric condition. Separation of these issues is, at times, difficult, if not impossible. The pathological somatic disorders in which fatigue occurs are easier to understand.

Learning outcomes

After studying this chapter, the student should be able to:
- explain the term fatigue;
- understand the role of fatigue and stress in society;
- identify the types of disorders in which fatigue plays a part, and give some illustrative examples.

2. Neurological and psychiatric disorders

Fatigue is common in certain neurological (and muscular) conditions. It is an anticipated symptom of disorders of the neuromuscular transmission, as is seen in *myasthenia gravis* or primary muscular disease (for instance, *muscular dystrophies*). In *multiple sclerosis*, a demyelinating disease of the central nervous system, fatigue is often the presenting symptom, and it may be present for some considerable period before other neurological signs can be detected. *Neuroses* and depressive psychoses are accompanied by severe fatigue. *Dementia* of any type frequently presents with fatigue.

3. Metabolic disorders

Conditions causing fatigue include diabetes mellitus, hypothyroidism, Addison's disease (adrenal cortex insufficiency), and panhypopituitarism (anterior pituitary deficiency). In *diabetes mellitus*, the metabolism of glucose, sodium, and potassium is abnormal. This compromises the normal sodium pump mechanism at cellular level and particularly affects the muscles. A comparable sodium and potassium metabolism defect exists in *Addison's disease*.
In *hypothyroidism,* the basal metabolic rate is reduced. In *anterior pituitary deficiency,* lack of several hormones leads to underactivity of the target endocrine glands such as the thyroid, adrenal cortex and gonads, with resulting metabolic induced fatigue. Potassium deficiency, for instance due to excess diuretic therapy or diarrhoea and vomiting, causes fatigue as does the excess calcium found in *hyperparathyroidism*, *sarcoidosis*, and sometimes in widespread *malignancy*.

In all of these conditions the underlying pathophysiology probably lies at neuromuscular or muscular cell metabolism level.

4. Malignant disease

Fatigue of an overwhelming nature is experienced by a very large number of patients with malignant diseases. The causes are not as yet well understood, but contributory factors including toxic effects from tumour metabolic products are recognised. Co-existent anaemia and malnutrition together with nausea and anorexia also lead to fatigue. In overt malignancy the patient is in a very stressful situation, a factor which is sometimes sadly neglected in this era of intrusive, 'high-tech' investigative based medicine.

Nevertheless, any patient complaining of fatigue, weight loss, and anorexia needs careful consideration if a treatable malignancy is not to be missed.

5. Oxygen deficiency

As metabolic processes require oxygen, and energy production is dependent on normal metabolism, fatigue will inevitably accompany any disorder which involves impaired oxygen absorption, transport, or utilisation. Previous chapters have described such situations in diseases of the heart or respiratory tract and in cases of anaemia.

6. Inflammation and infection

In many inflammatory disorders there is an increased metabolic rate. Some infective organisms produce systemic toxins. Specific toxins target the neuromuscular systems. All of these are causes of fatigue. Less easily understood is the profound fatigue sometimes experienced after an infectious disease (especially a viral one). So called *myalgic encephalomyelitis (M.E.)* may be such a condition. The percentage of sufferers of this condition, if it exists, who have a significant underlying psychological component is not quantified.

7. Summary

There are many and various causes of fatigue, and it is not a complaint to be lightly dismissed. It is often the prodromal presenting complaint in a patient with a serious but still occult underlying illness. The possibility of the presence of this hidden pathology must always be considered.

Urinalysis

24

1. Introduction

Large amounts of soluble waste products are eliminated daily in the urine. These waste products include urea, uric acid, protein degradation products, excess electrolytes and water. The normal if complex homeostatic mechanisms control the various concentrations of these compounds in the urine. It is therefore evident that systemic disorders will, to a greater or lesser extent, be expressed in abnormal urine production either by volume or by composition. It goes almost without saying that urinalysis is of paramount importance in investigating renal disorders.

Learning outcomes

After studying this chapter, the student should be able to:
- explain micturition and the origins of incontinence;
- understand the significance of the quantity, colour, and specific gravity of the urine;
- understand the significance of the presence of sugar, protein, urobilin, bilirubin, and acetone in the urine;
- understand the significance of microscopic examination of the urinary sediment;
- list and describe the data obtained from urine examination.

2. Micturition

Micturition is controlled by complex interacting neuro-muscular actions involving, in a healthy individual, both automatic and voluntary neural pathways.

Urine produced in the kidneys is conveyed by the ureters to the bladder. This increasing volume of urine stretches the muscular wall of the bladder, and when approximately 400ml of urine are present the degree of bladder-wall stretching triggers involuntary contraction, with subsequent bladder emptying unless there is voluntary inhibition. Impulses from the higher centres in the cerebral cortex can prevent bladder emptying if it is deemed inappropriate by the individual. Shortly after such voluntary inhibition of micturition, the bladder's muscular wall further relaxes to provide temporary relief from the sensations of need to pass urine. However, the more the bladder fills with urine, the more the pressure increases and the urge to void urine again becomes conscious. This process may repeat several times until the individual consciously relieves the situation and empties the bladder.

Injury to the nerve pathways which control this mechanism leads to inability to voluntarily control micturition.
This is seen for example in patients suffering from multiple sclerosis, dementia, Parkinson's disease, and as the result of a cerebral vascular accident. Loss of higher control allows the bladder to proceed with reflex contraction when the intra-vesical pressure exceeds the trigger threshold. The patient as a result suffers incontinence. Local pelvic pathology may lead to sphincter opening at less than optimal pressure, causing *stress incontinence*. Thus, patients with genital prolapse, post-partum urethral damage, urinary tract infection or even constipation may experience stress incontinence if the bladder pressure transiently rises as a result of such activities as sneezing, coughing, laughing, or lifting. In elderly males with an enlarged prostate gland at the bladder neck, the internal sphincter may have impaired function leading to dribbling as a result of pressure in a chronically distended bladder – this is called *overflow incontinence*. The elderly, infirm or mentally impaired may demonstrate *urgency incontinence* if a toilet cannot be physically reached in time. *Nocturia*, or the passage of urine at night, is usually pathological unless it is due to excess intake of fluid, often alcoholic and late in the evening. In all other instances nocturia indicates some abnormal condition, such as heart failure, prostate enlargement, renal disease and many others.

3. Volume of urinary output

The average normal daily output of urine is 1.5 litres. When the volume of urine exceeds 2.0 litres in 24 hours the term *polyuria* is applied. This arises if large amounts of soluble waste products need to be eliminated in conditions such as hypercalcaemia in hyperparathyroidism, or if blood degradation products require clearing (after gastro-intestinal haemorrhage), or during states of excess protein breakdown.
Lesions of the posterior pituitary, leading to a deficiency of anti-diuretic hormone, cause large volumes of urine of low specific gravity to be produced. This condition is known as diabetes insipidus. Conditions producing excess glucose or electrolytes, such as diabetes mellitus, give rise to polyuria as the renal mechanisms attempt to control the normal levels in the blood.

The volume of urine produced may also be reduced. When less than 500ml urine is produced in a 24-hour period *oliguria* is present. *Anuria* implies no effective urine production. Oliguria and anuria are significant and serious clinical signs because the body will be unable to excrete soluble waste products in adequate amounts.

There are various causes of oliguria. Firstly, the intake of fluid may be greatly decreased due to environmental lack of water or to immobility or infirmity preventing normal drinking. Vomiting, diarrhoea, and excess sweating can lead to dehydration and subsequent oliguria or even anuria. In states of clinical shock the renal blood flow is often seriously impaired, with consequent reduced renal output. Congestive cardiac failure similarly leads to decreased renal function accompanied by an accummulation of fluid in the tissues (oedema). Many renal diseases, including acute glomerulonephritis, diabetic glomerulosclerosis, renal artery arteriosclerosis and advanced pyelonephritis adversely affect kidney function. Urine production may be normal but the volume passed may be low in patients with an obstructive lesion in the lower urinary tract. An enlarged prostate is the commonest and best known example.

In all of the above cases, levels of toxic metabolites increase in the body. Uraemia is the most obvious abnormal measured parameter in these conditions. It is important that in these cases accurate records of all fluid intake and output is made if fluid balance is to be corrected and maintained.

4. Colour of the urine

Most of the metabolic products present in the urine are colourless in solution (mineral salts, urea, uric acid, and creatinine). Urine is normally pale yellow due to the presence of metabolic products from the breakdown of bilirubin. In a healthy individual, the colour will darken if the volume of urine is relatively reduced as a result of lowered fluid intake and, conversely, it will lighten when fluid intake increases. The urine is almost colourless after a heavy bout of social drinking.

When the quantity of urobilin increases, as in hepatic disease or haemolysis, the urine colour darkens almost to a mahogany hue. In biliary obstruction, whether due to stones, inflammation, or neoplasm, the urine darkens as a result of the presence of large amounts of bilirubin. This urine has a frothy appearance when shaken, as a result of the bile salts content.

A red coloration may result from the presence of blood. Blood in the urine is referred to as *haematuria*, and it may change the colour from hardly visible pink to an almost opaque dark red, depending on the amount. When visible to the naked eye, the term macroscopic haematuria pertains, and when the red blood cells in the urine are only detectable by microscopy, it is referred to as microscopic haematuria (Figure 24.1).

Figure 24.1
Macroscopic haematuria

Haemoglobin released into the urine from lysed erythrocytes (haemoglobinuria) also colours the urine red, but of a transparent, not turbid, nature. Foods including beetroot can colour urine red, as can some medicinal compounds, and phenolphthalein produces a red coloration in alkaline but not in acidic urine.

5. Specific gravity of the urine

The concentration or specific gravity of urine is determined by the number of particles dissolved in it. Normal urine has a specific gravity of 1.010 to 1.025. In physiological relative oliguria there is concentrated urine and in the presence of normally functioning kidneys this has a high specific gravity. Decreased volume of urine with a low specific gravity indicates a renal inability to concentrate the dissolved electrolytes. In cases of polyuria, the specific gravity will only be increased if large amounts of relatively heavy molecules such as glucose or dissolved protein are present.

6. Glucose in the urine

The presence of the sugar glucose in the urine is referred to as *glucosuria* or *glycosuria*. As glucose molecules can only be excreted in solution, the greater the amount of glucose excreted the larger the volume of urine which is produced. This inevitably results in polyuria. The urinary specific gravity rises due to the dissolved glucose.

'Near patient testing' by means of specific glucose-sensitive test strips offers confirmation of the presence of glucose (Figure 24.2). Test strip assay of glucosuria is reliable if carefully performed as a semi-quantitative assessment. The detection of the presence of glucose in the urine is often the first indication of diabetes mellitus.

Normally, the glucose excreted by the glomeruli is fully reabsorbed in the renal tubules, and the urine of a healthy person therefore contains no glucose. When the glucose level in the blood is elevated, as in diabetes mellitus, the glomerular excretion exceeds the capacity of tubular re-

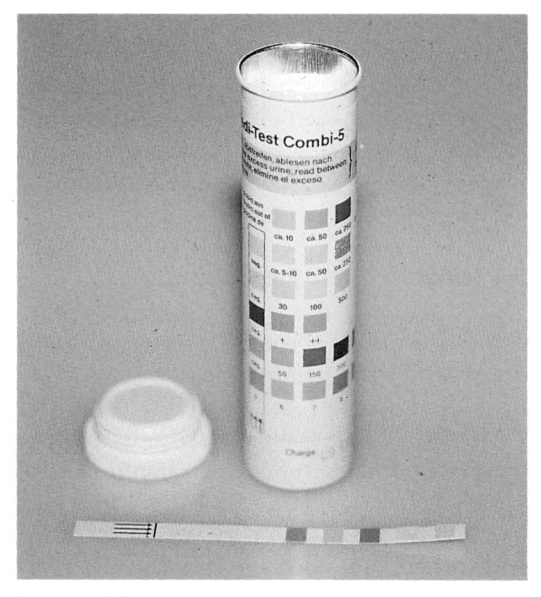

Figure 24.2
Urine testing with 'dipsticks'

absorption. In this situation, glucose is present in the urine.

Temporary glycosuria may appear in pancreatitis, when temporary insulin deficiency occurs. In pregnancy the renal threshold may be exceeded, and this can also happen occasionally after gross overloads of glucose in the diet. In diabetes mellitus, carbohydrate metabolism is so impaired that catabolic activity switches to fat metabolism for energy production. Metabolising fat for energy is less efficient than metabolising carbohydrates, so abnormal degradation products appear in the urine. These include fatty acids, acetoacetic acids, and ketones. The presence of these fat metabolic products in the urine always indicates severe diabetes requiring immediate treatment. After a period of fasting, or in starvation, similar ketone bodies appear in the urine (*acetonuria* or *ketonuria*). Acetone is also excreted in the breath with a characteristic odour of sour apples.

7. Protein in the urine

In health, no protein passes from the renal tubules so no significant amounts can be detected in the urine. (Care must be taken to avoid contamination of the specimen with proteinaceous material such as vaginal discharge.) In certain renal disorders, however, the glomeruli are permeable to proteins, which are not reabsorbed in the renal tubules and which therefore appear in the urine. This is known as *proteinuria*. Testing with diagnostic dipsticks can be useful and, if carefully controlled, semi-quantitative (Figure 24.3).

8. Hormones in the urine

Numerous hormones and their metabolites are excreted in the urine. Measurement of these may be useful in the investigation of endocrine abnormalities and for monitoring feto-placental function in pregnancy. Trophoblastic B-HCG is present in the urine in pregnancy and forms the basis of most pregnancy tests. Hormone markers in some malignant tumours can also be assessed from urinalysis and can provide a prognostic indication.

9. Urinary sediment

Some urinary constituents are not wholly soluble. When they are present in high concentrations, these substances precipitate.
Crystal formation of uric acid, phosphates and oxalates may, for example, occur. In renal

Figure 24.3
Dipstick examination of urine

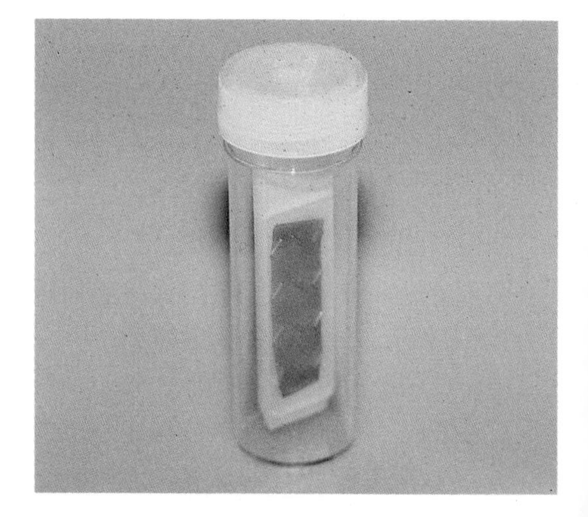

disorders, protein and erythrocyte casts may form. Examination of urine after centrifugation and preparation for microscopy will reveal these casts. Infection in the urinary tract will often result in the presence of bacteria in the urine (*bacteriuria*). Culture of the urine by the microbiology department will specify the organism and give an indication of antibiotic sensitivity. Sterile pyuria may indicate urinary tract tuberculosis.

10. Summary

Involuntary passage of urine is known as incontinence, and several types have been described. Changes in urinary volume may indicate a variety of metabolic disorders. Careful recording of urinary output and fluid balance is essential in many clinical situations. The colour, smell and specific gravity of urine and the presence of glucose, protein, blood casts and other sediments provide valuable information in the diagnosis of systemic and renal disease.

25 Blood

1. Introduction

To the lay person, blood is a red, sticky, slightly salty liquid only encountered directly as a result of minor personal trauma. Blood flows through virtually all body systems and tissues. It conveys oxygen and a variety of other substances (mainly nutrients) necessary for normal metabolism, to the tissues, and carries carbon dioxide and other waste metabolites back from the tissues for excretion. In effect, blood acts as the energy supplier. It provides the transport medium for many other regulating substances such as enzymes and hormones, and it thus provides a communication network between various distant bodily systems. The composition of the blood, therefore, reflects accurately the levels of efficiency and activity at which many body systems are working.

Pathological processes manifest themselves by a change in the composition of the blood and its various components. Examination of the blood can therefore provide much information relating to many disease processes.

Learning outcomes

After studying this chapter, the student should be able to:
- describe the methods of obtaining capillary, venous, and arterial blood samples;
- list and describe the cellular components of the blood and some conditions which may affect them;
- list and describe the main components dissolved in the serum and the conditions which can alter their concentration;
- identify some haematological investigations, including blood films, protein analysis, enzyme assay and coagulation studies;
- identify the disease processes which alter the outcomes of these investigations.

2. Capillary blood sampling

A very simple and safe method of obtaining a blood specimen is by capillary sampling (Figure 25.1). A lancet of regulated penetration depth is stabbed into the finger pulp or ear lobe (in young children the heel is used). The escaping blood is collected into capillary tubes for testing.

Samples thus obtained are suitable for haemoglobin level measurement, red and white cell counts, platelet count, reticulocyte count, and (if required) blood gas analysis. This method is also a suitable way of measuring 'bleeding time'.

The amount of haemoglobin, together with the number of *erythrocytes* (red blood cells), provides assessment of any anaemia. Initial indication of polycythaemia can also be made. *Reticulocytes* (immature red cells still possessing nuclear material) can be recognised and counted, and their presence is an indication of active red cell manufacture (*erythropoesis*). In certain types of anaemia, especially after blood loss, the number of reticulocytes circulating in the peripheral blood is significantly increased. Many inflammatory states or infections lead to an increase in circulating *leucocytes* (white blood cells). Malignant, immature leucocytes are detectable in cases of leukaemia, and a decrease in the number of leucocytes is found in certain marrow disorders as a result of toxic depression (sometimes due to medication) and where the marrow cavity is infiltrated by malignant tumour. This decrease in numbers of leucocytes is called *leucopenia*. Total absence of leucocytes is called *agranulocytosis*. Both these conditions are potentially serious or even fatal as the body's defence mechanisms are compromised.

A blood film is easily prepared by smearing a drop of blood on a microscope slide and staining it (Figure 25.1 and 25.2). This enables the blood cells to be examined in detail and the proportions of the different cells to be determined (differential count). An assessment can be made of the presence of abnormal cells, variations in size and shape, abnormal clumping of cells (rouleaux formation), and recognition of parasites such as malaria. Platelet numbers and measurement of 'bleeding time' can readily be undertaken from a capillary 'stab'.

Blood coagulation is important in a variety of conditions including marrow depression, liver disorders, where patients are on anti-coagulation therapy and many others. Coagulation studies are needed before certain invasive procedures are carried out; percutaneous liver biopsy is one important example. Most of the above parameters are now readily measurable by automated processors to very high levels of accuracy.

Blood gas analysis (see also arterial blood sampling) reflects respiratory and metabolic function. The concepts of relative acidosis and alkalosis are worthy of brief description.

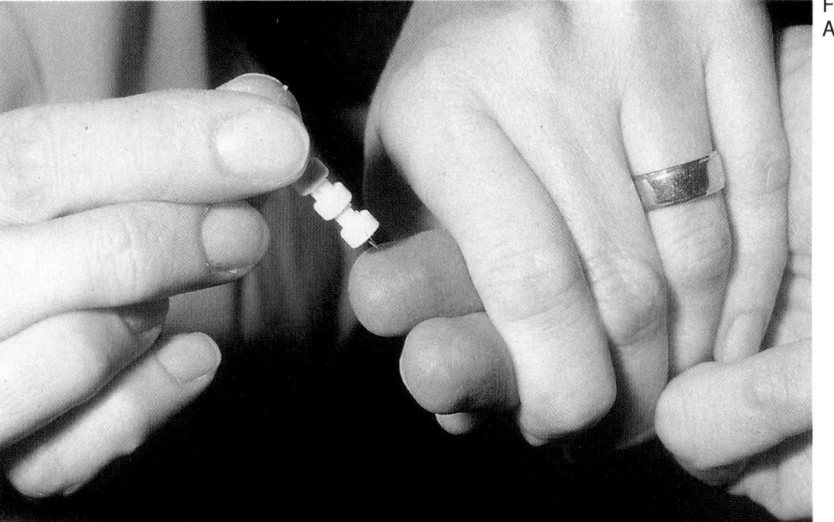

Figure 25.1
A finger prick

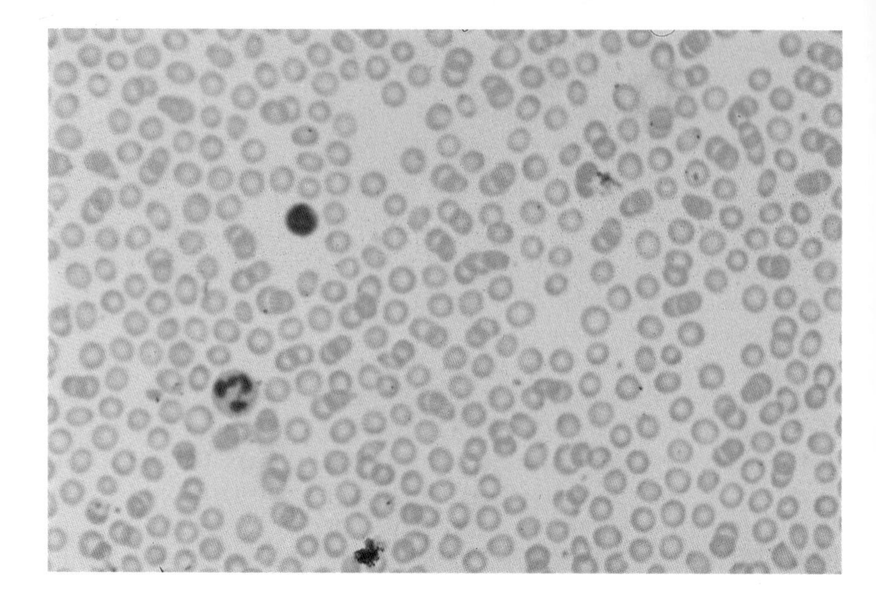
Figure 25.2
Stained blood film

Acidosis implies a shift of the normal acid/alkali balance towards the acid (i.e. an increase in acidity and fall in pH). The symbol pH relates to the hydrogen ion concentration in a solution, with neutral having a value of pH 7 (i.e. pure water), values below pH 7 indicating acidity and values above pH 7 indicating alkalinity.

Alkalosis implies the opposite, namely a disturbance of the balance towards increased alkalinity with increased pH.

In impaired respiratory function there is oxygen deficiency together with increased acidity from retained carbon dioxide leading to respiratory acidosis. Shock or other metabolic disorders cause acid accumulation in the body, leading to metabolic acidosis. On the other hand when acid is lost, for instance due to persistent vomiting, metabolic alkalosis results. When loss of acid is a consequence of increased expiration of carbon dioxide, due for example to hyperventilation, respiratory alkalosis exists.

3. Venous blood sampling

In order to carry out many laboratory investigations, larger volumes of blood are required than can be obtained from capillary sampling. Venous samples from a suitable superficial vein are then obtained (Figure 25.3).

Many veins are located superficially and are therefore easily accessible for venepuncture (i.e. puncture of a vein to remove blood samples – also known as phlebotomy). Access to veins is more difficult in children and the elderly. The elderly often have prominent veins which are mechanically difficult to enter, often resulting in substantial local haemorrhage and bruising. Special skin preparation is needed if the blood sample is to be cultured for pathogens in cases of suspected septicaemia.

Blood samples can be prepared to divide the *serum* from the cells. Serum is the liquid which separates spontaneously from coagulated blood. As part of this process, soluble fibrinogen is converted to fibrin which forms a dense protein mesh and traps the red cells to form a clot.

Blood which has been treated with an anticoagulant does not clot, but the red cells can be separated out by centrifugation. The supernatation fluid still contains fibrinogen and constitutes plasma.

The ratio of red cell volume to plasma volume is called the *haematocrit*. The larger the overall volume of the cellular component, the higher the observed haematocrit. The haematocrit obviously increases during dehydration and decreases during haemodilution, and it thus

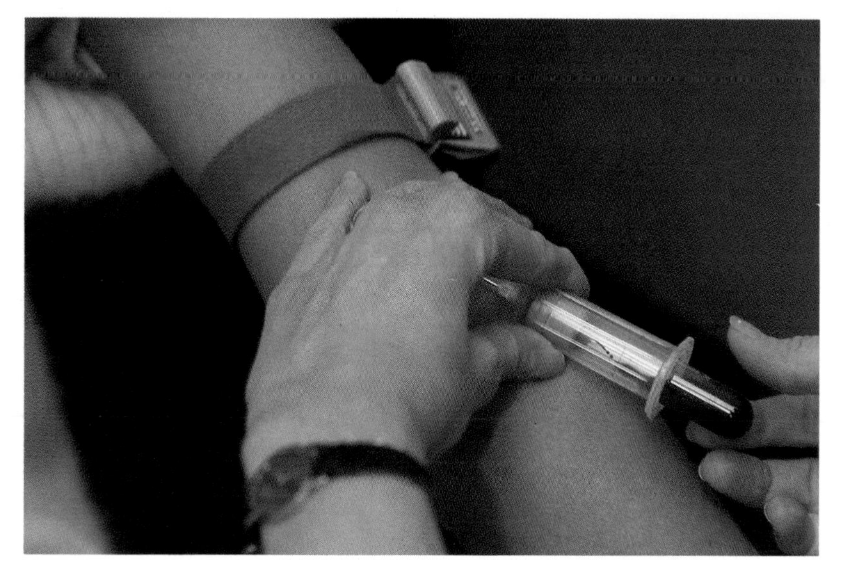

Figure 25.3
Venepuncture

provides a useful measurement of fluid balance. The normal value is in the range 0.45 – 0.50.

The *erythrocyte sedimentation rate* (ESR) measures the rate of settling of a tube of blood containing an anticoagulant (Figure 25.4). The ESR, although a non–specific indicator, has historic usage. The rate is increased in many non–related conditions such as infection, malignancy, and auto–immune disorders. The normal range lies between 5 to 10 mm in the first hour and is normally slightly higher in females. A more subtle measurement is calculation of red cell viscosity.

Numerous biochemical measurements can be undertaken on a blood sample, and a few of the more important ones are referred to here (see the appendix on page 147 for a list of normal values).

Urea is one of the most important waste products of protein metabolism. It is manufactured in the liver and excreted by the kidneys. An increase in blood urea results from either increased tissue breakdown (catabolism) or impaired renal function. *Creatinine* levels are a more specific indication of renal function.

The level of *bilirubin* is increased in both biliary obstruction and haemolysis. The extent of the elevation of the bilirubin level parallels the severity of the underlying condition.

Enzymes of several types are regularly measured in blood samples. Enzymes are not infrequently present in different structural forms, and they are then called *iso-enzymes*. It is possible to measure two types of phosphatase enzyme in the blood. *Alkaline phosphatase*, which shows activity in an alkaline environment, is mainly found in bone and liver. *Acid phosphatase* on the other hand, which is active in an acid environment, is produced primarily in the prostate.
An elevated alkaline phosphatase suggests liver or intrinsic bone pathology, whereas elevated acid phosphatase suggests metastatic prostatic carcinoma.

There are likewise two common types of *transaminase* (now called *aminotransferase*). AST (serum aspartate aminotransferase), formerly referred to as SGOT (serum glutamic oxaloacetic transaminase) is found especially in cells of the myocardium and is much elevated after a myocardial infarction. The SGPT (serum glutamic pyruvic transaminase) is mainly present in liver cells and is elevated in hepatic diseases such as hepatitis. The CPK (creatine phosphokinase) level is elevated in disorders of the skeletal and cardiac muscle. Iso-enzyme studies enable separation and measurement of CPK in the early stages of a myocardial infarction, and these are most useful in assessing degree of myocardial damage.

Elevated gamma glutamyl transpeptidase is found in biliary stasis and notably in alcohol abuse. LDH (serum lactic dehydrogenase) consists of five iso-enzyme varieties and is elevated in myocardial infarction, liver disease, haemolysis, and often in malignant disease. Serum amylase elevation suggests pancreatitis. Analysis of the crude protein level in the blood is only of limited use in assessing basic nutrition. The albumin level is the better guide to overall nutrition. Comparison of albumin and globulin levels, ratios, and globulin components is useful in the diagnosis of immune disorders and certain monoclonal malignancies such as multiple myeloma.

The *blood glucose level* varies between tightly maintained limits, normally modified by food intake. A subnormal concentration of blood glucose is called *hypoglycaemia*, a greater than normal blood glucose concentration is called *hyperglycaemia*. The level of blood glucose in diabetes diagnosis and in diabetics receiving insulin is crucial. *Lipid levels* (possibly associated with coronary heart disease) are currently in fashion. Whether a knowledge of cholesterol and triglyceride levels is of significant importance to most individuals (excepting the very small minority with familial hypersterolaemia) is not yet properly evaluated.

In anaemia, *iron* levels are low due to malnutrition and blood loss or, sometimes, to chronic infection or states of increased need such as pregnancy. A study of iron metabolism however is more complex, involving assessment of uptake, utilisation, storage and binding capacity. Measurement of the *potassium* and *sodium* levels are undertaken routinely and provide a basis for assessment of salt and fluid balance. This is especially true in cases of dehydration, renal disease, liver disease, adrenal disease and shock. *Calcium*, which is needed for bone metabolism and for muscular action, is present in the blood in two principle forms; partially ionised and in solution, and partially protein bound. The blood level is controlled by the parathyroid gland and the availability of vitamin D.

Elevated plasma calcium is found in lesions of the parathyroid, bone disorders and especially in malignant lesions and other conditions, including sarcoidosis. A low plasma calcium may indicate chronic renal failure. Factors controlling coagulation can be measured. PTT (partial thromboplastin time) is a useful guide to the efficacy of anticoagulant therapy but may be increased by various disease processes involving the liver and, to a lesser extent, the kidneys.

4. Arterial blood sampling

The femoral, radial and brachial arteries are the common sites for arterial blood sampling (Figure 25.5).

Figure 25.4
Normal and increased E.S.R.

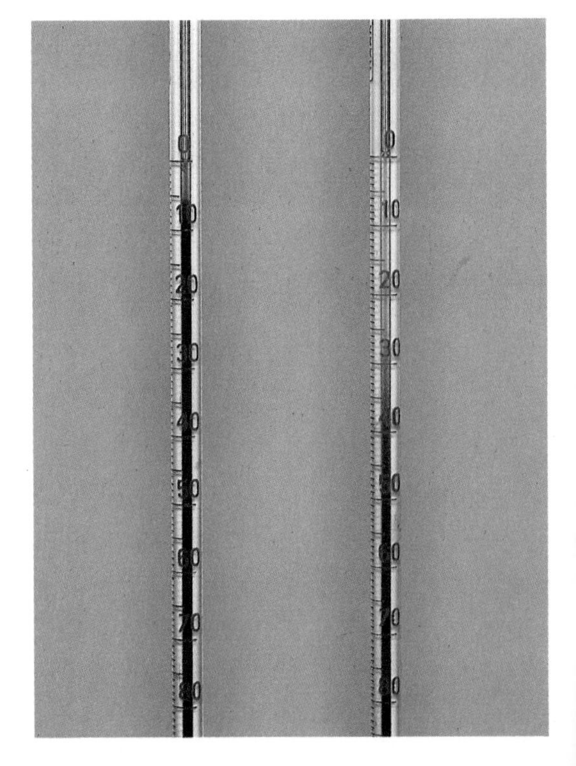

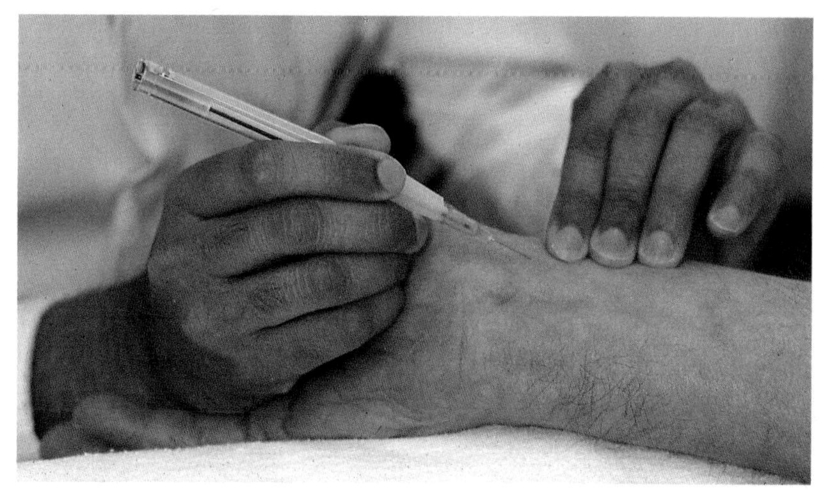

Figure 25.5
Arterial puncture of the radial artery

Special techniques are required to seal the blood if arterial blood gases are to be measured. After sampling, pressure on the artery will prevent local leakage and bruising. Cardiac catheter samples are undertaken for specific measurement of cardiac and pulmonary function, a long catheter being inserted via a peripheral artery or vein as required.

Blood ammonia is measured most reliably from arterial blood. A healthy liver converts the ammonia produced in the intestines into urea, which is subsequently excreted by the kidneys. The level is elevated in hepatic disease.

5. Normal levels

A summary of normal levels is provided in the appendix on page 147.

6. Summary

Capillary blood can be used to measure haematological indices including haemoglobin and cell counts. As capillary blood in many ways approximates to arterial blood it can be used for some blood gas values. Venous samples are taken for ESR measurements and for most routine biochemical tests. The presence of iso-enzymes and the uses made of their measurement have been discussed.

The composition of the blood accurately reflects the function of many body activities. Impaired function is often indicated by an increase or decrease of a specific blood constituent. Blood samples can be obtained from capillaries, veins, and arteries.

The results obtained from blood tests must always be compared with the normal ranges of the testing laboratory.

26 Faeces

1. Introduction

The characteristics and composition of faeces reflect in many ways the functioning of the gastro-intestinal tract. Various disorders of the gastro-intestinal tract leave traces in the faeces. Many of these abnormalities are visible to the naked eye, but laboratory examination is often necessary to detect more subtle or microscopic features.

Learning outcomes

After studying this chapter, the student should have sufficient knowledge and understanding of:
- patterns of defecation, and faecal composition and variations;
- the causes of diarrhoea;
- the causes of constipation;
- faecal incontinence.

2. Defecation

Digestion of food products in the stomach and small intestines leaves behind roughage, microorganism debris and water, together with discarded intestinal epithelial cells. Peristalsis moves this mixture further down the colon where the remaining water is reabsorbed. The residue in the sigmoid colon constitutes the faeces. In the sigmoid colon the increasing pressure causes involuntary relaxation of the internal anal sphincter, and under normal voluntary control the external anal sphincter then relaxes and defecation occurs.

In inappropriate social situations, voluntary control delays defecation until there is a socially acceptable opportunity, until which time the faeces are moved back above the internal sphincter.

The *frequency* of defecation is very much an individually determined habit. Some individuals normally defecate three times a day with no gastro-intestinal abnormality, whereas others may only defecate once every two or three days. The majority of people eating a European diet defecate once per day, but a more frequent pattern does not necessarily indicate diarrhoea as long as faecal constituency is normal. Similarly, defecation only once every three days may not be a sign of constipation if this is the usual pattern for the individual. So-called *fake urgency* after defecation is almost always pathological and is often an indication of a low rectal neoplasm.

3. Characteristics of faeces

a. Consistency

The consistency of the faeces reflects gastrointestinal function, whether constipated or diarrhoeal. Care however must be exercised as great dietary variation occurs. The amount of roughage, especially in vegetarians, greatly alters the faecal texture.

b. Odour

The odour of faeces produced from a European diet is recognisable with little difficulty. Large amounts of garlic, onions or gas-producing food produce a typical smell.
Malabsorption of proteins produces a marked sulphurous smell and carbohydrate digestion abnormalities can produce a sour odour caused by the presence of large amounts of carbohydrates. This also occurs in problems of fat digestion resulting from the presence of excess fatty acids. Black, tarry faeces containing altered blood (melaena) have a characteristic metallic smell.

c. Colour

The colour of faeces is mainly determined by the waste products derived from bilirubin in the intestines, especially the urobilin and stercobilin, but it is also affected by the nature of the food consumed. The normal colour is dark brown.

In biliary obstruction the normal biliary colorants are largely absent and the faeces are consequently pale.

Care is needed here to distinguish the above from faeces in steatorrhoea, which contain large amounts of partially digested fat materials. In steatorrhoea the faeces are more ash-grey, and they often float. Steatorrhoea (Figure 26.1) is seen in chronic pancreatic insufficiency and in cases of tropical sprue. Low intestinal haemorrhage colours the faeces bright red, as opposed to black, in upper gut bleeding. Substances such as beetroot must be remembered, however, as alternative causes of red faeces.

Figure 26.1
Bloodstained and grey-white 'putty faeces' – in steatorrhoea

d. Presence of abnormal substances

Blood in the faeces is always abnormal. If the blood is red and clearly visible the site of haemorrhage is likely to be in the colon or lower portion of the small intestine. The causes are numerous and may include enteritis, such as typhoid, dysentery, and other intestinal infections. Inflammatory bowel diseases such as ulcerative colitis or Crohn's disease often result in haemorrhage. Tumours both benign and malignant frequently bleed into the gut, carcinoma being the most important. Anal haemorrhoids often cause a brisk haemorrhage at time of defecation. A childhood bowel intussusception often presents with rectal bleeding.

When the faeces contain large amounts of mucus or pus or both, an inflammatory process is present in the intestines. These conditions include infective diarrhoeas and inflammatory bowel disease, including Crohn's disease and ulcerative colitis.

Stool culture can isolate infective micro-organisms, and these are discussed later in this chapter. Tapeworms, roundworms, threadworms, and flukes, although more common in a tropical environment, are found from time to time in Europe, as world travel becomes more commonplace. Threadworms, as a reflection of hygiene, are still very common in children. Recognisable food material is rarely seen in faeces, but indigestible foreign bodies such as coins or jewellery, often swallowed by children, can be rescued after passage through the intestines.

4. Anus

Anal pathologies are usually readily visible on clinical examination. Haemorrhoids (piles), fissures, fistulae, perianal abscesses, and benign skin tags are common. They may be accompanied by pain, itching, bleeding, excoriation, and embarrassment. Low rectal carcinomas and anal tumours are accessible to digital examination. Rarer tumours such as malignant melanomata may mimic haemorrhoids. Anal tumours may cause faecal incontinence or urgency. Prolapse of rectal mucosa in the elderly is common.

5. Faecal incontinence

Disorders of the internal or external anal sphincter may lead to faecal incontinence, and the condition is also common in dementia or brain damage after cerebro-vascular accident. Local anal lesions including haemorrhoids, fissures, tumour, or trauma may also be causative. In certain severe diarrhoeal illnesses faecal incontinence may be a complicating factor. Spurious 'diarrhoea' in the elderly masks underlying constipation. In this condition, hard faeces, unable to be expelled, remain in the upper rectum and they are bypassed by more liquid faeces together with mucus which escape through a somewhat patulous anus. Suitable diet and controlled use of laxatives for a strictly limited course are usually curative.

6. Diarrhoea

Diarrhoea always indicates some gastro-intestinal malfunction, usually more inconvenient than serious if of short duration. In general, it is accompanied by abdominal cramps and is rapidly exhausting. In tropical countries diarrhoeal illnesses in children are all too often fatal due to severe loss of water and salt. Simple salt and water oral replacement is life saving in these circumstances, and the World Health Organisation has a vigorous campaign to encourage use of this simple management. Oral rehydration fluid made up from 20 gr Glucose, 2.5 gr NaCl, 2.5 gr NaHCO3 and 1.5 gr KCl per litre of solution in water is cheap and highly efficient.

The causes of diarrhoea are many, and only a few of the more important are illustrated here.

a. Gastroenteritis

Virtually everybody has suffered a bout of gastroenteritis. The sufferer experiences diarrhoea, abdominal cramps, nausea, vomiting, and general weakness. Viral infection is a common cause of the problem. Infection is acquired, often due to careless hygiene, by faecal-oral transmission. The condition is usually short lasting and self limiting without significant treatment.

b. Bacterial enteritis

Many bacteria including staphylococci, streptococci, bacilli (such as salmonella typhi), vibrios (such as cholera) and shigella (bacillary dysenteries) are implicated. The clinical condition varies from mild to life threatening, and the patient's previous state of health often determines the outcome. Stool culture will provide a specific microbiological diagnosis. Appropriate rehydration is more important than antibiotic therapy. Travellers can prevent many such intestinal infections by scrupulous attention to cleanliness and avoidance of potentially contaminated foods.

c. Toxic enteritis

Some food materials, or elements in them can be toxic to some individuals but not others. Diarrhoea is not rare after consuming unripe fruits or large amounts of rhubarb or raw vegetables. Food 'going off' may contain bacterial toxins which lead to gastroenteritis.

Hypersensitive reaction to certain foods may cause diarrhoea in sensitive individuals. Common culprits include crabs, prawns and lobsters. In these cases, the food need not be bad. Certain mushrooms are poisonous (amanita phylloides) and may in fact prove fatal, and berries of several types are irritant and cause diarrhoea.

Laxative abuse is a frequent cause of diarrhoea. Chronic laxative use causes diarrhoea often followed by bouts of constipation setting up a vicious cycle. Frequent laxative ingestion may lead to dehydration and salt depletion.

d. Intestinal parasites

Threadworms are very common in children as a result of faecal-oral transmission of the ova or parasites. In more tropical climates intestinal worms are a major health hazard. These include tapeworms (such as *Taenia saginatum*, or beef tapeworm) (Figure 26.2), roundworms (*Ascares*), ancylostomiasis and many others. Examination of the faeces followed by specific medication is required to eliminate such parasites.

e. Disturbed digestive function

Impaired digestion, resulting from poorly functioning digestive enzymes, or their absence, causes undigested food to pass to the colon with subsequent diarrhoea. Diseases of the intestine and the pancreas are common causes and include chronic pancreatitis and sometimes inflammatory bowel disease such as Crohn's disease.

f. Malabsorption

Some conditions result in abnormal absorption of digested food. Coeliac disease provides a good model. Hypersensitivity to ingested gluten in the diet (a major constituent in cereals) leads to inflammation and atrophy of the villi of the upper small intestine with malabsorption as a direct consequence. Treatment involves use of a gluten-free diet. In many non-specific inflammatory conditions of the small intestine, temporary malabsorption may be present.

g. Tumours

Although tumours are not a major cause of diarrhoea, their presence often causes an alteration in bowel habit and transient diarrhoea may alternate with episodes of constipation.

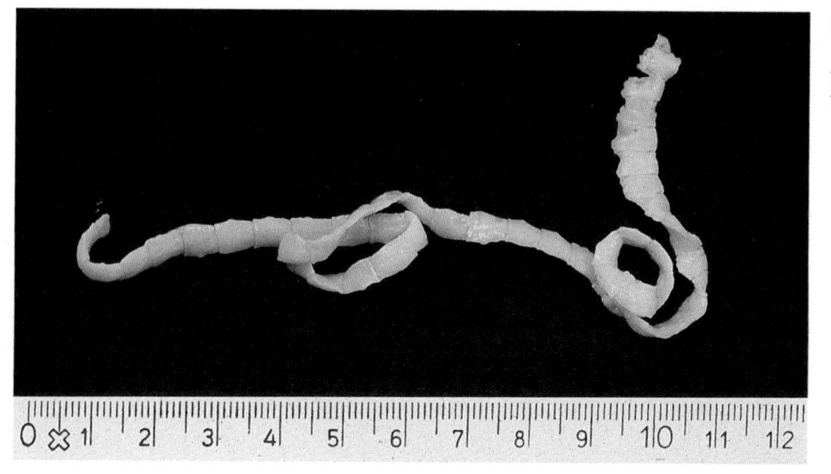

Figure 26.2
Taenia saginata or beef tapeworm (scolex or head to the left)

7. Constipation

Constipation is determined by the passage of abnormally hard faeces, and not by frequency of defecation. Constipated faeces are painful to pass and the mechanical difficulty tends to exacerbate the problem, especially in the elderly.

Apart from general infirmity or an inappropriate diet, some other causes of constipation are worthy of brief mention.

In Hirschsprung's disease there is a congenital lack of normal innervation in the lower colon. Local obstructive lesions of the lower colon, rectum, and anus (especially tumours) result in constipation. Any lesion causing pain on defecation will tend to result in avoidance of defecation and thus constipation. Simple dietary, social, or stress factors should always be considered before embarking on intrusive investigations.

8. Summary

Defecation or bowel movement is the elimination of faeces, and normally occurs under voluntary control. Involuntary passage of faeces results in faecal incontinence. The frequency of defecation is an individually determined habit. Examination of faeces to observe consistency, colour, and the presence of abnormal contents provides useful diagnostic information. The causes of diarrhoea have been discussed, as have the causes of constipation, and the uses and abuses of laxatives have been considered.

Sputum

27

1. Introduction

The airways from the nostrils to the bronchi contain numerous mucus-producing cells, together with a lining of other epithelial cells equipped with small hair-like projections (cilia). The mucus produced is a protective mechanism against dehydration and sudden temperature change. The cilia move in such a way as to waft the mucus upwards in the respiratory tract (the mucociliary mechanism). Many conditions adversely affect mucus production and damage the mucociliary mechanism.

Learning outcomes

After studying this chapter, the student should be able to:
- describe the mechanism and origin of coughing;
- identify when coughing is abnormal;
- identify and describe abnormal sputum production and its causes.

2. Coughing

Increased production of mucus or impaired ciliary motility may cause an accumulation of mucus in the airways, and this is expelled as sputum by means of the coughing reflex. After a deep inspiration, the cough drives the contents of the airways in the direction of the throat and mouth by means of forced expiration. In healthy individuals coughing occurs involuntarily to clear and clean the air passages. In the presence of inflammation, infection, irritation due to hypersensitivity, or tumour, the coughing becomes frequent and annoying.

Coughing may or may not be accompanied by expectoration of mucus. A cough which produces mucus is said to be productive, a dry cough non-productive. A *productive cough* indicates excess mucus production and is commonly seen in chronic bronchitis. It is also noted in cases of pneumonia, pulmonary oedema and sometimes in lung cancer. A *non-productive cough* indicates irritation rather than excess mucus production. This is seen if an irritant is inhaled, sometimes if there is an early developing tumour, and often in pleurisy.

3. Production of sputum

Examination of the sputum may provide valuable information about underlying respiratory disease (Figures 27.1 and 27.2).

Mucoid sputum is relatively thin, mucoid and transparent. It is characteristic of patients with non-infected chronic bronchitis. In asthma the sputum is thicker, viscous and coloured grey.

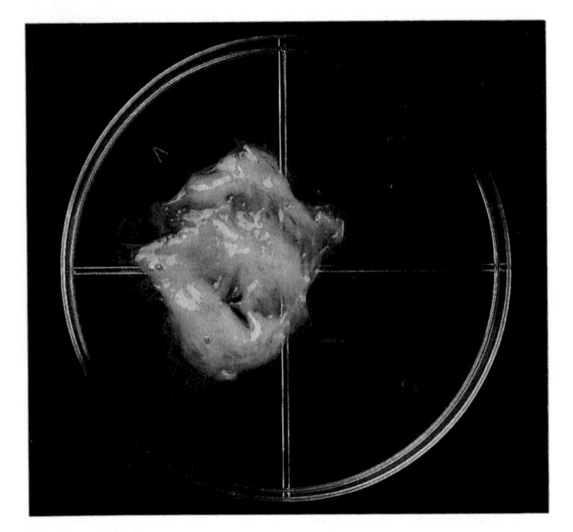

Figure 27.1
Thick, viscous grey sputum

Asthmatics experience difficulty in expectorating sputum.
In an inflammatory or infective condition of the respiratory tract the sputum becomes purulent and usually contains many leucocytes and dead bacteria. Purulent sputum is green or yellow.

Blood in the sputum (haemoptysis) is a cause for anxiety to the patient and requires investigation. Haemoptysis is seen in bronchitis, pneumonia, bronchiectasis, and malignant disease (especially bronchial carcinoma). Historically, in western Europe and in many other parts of the world it is still the presenting sign of pulmonary tuberculosis.

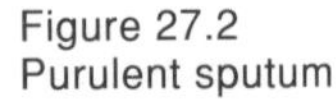

Figure 27.2
Purulent sputum

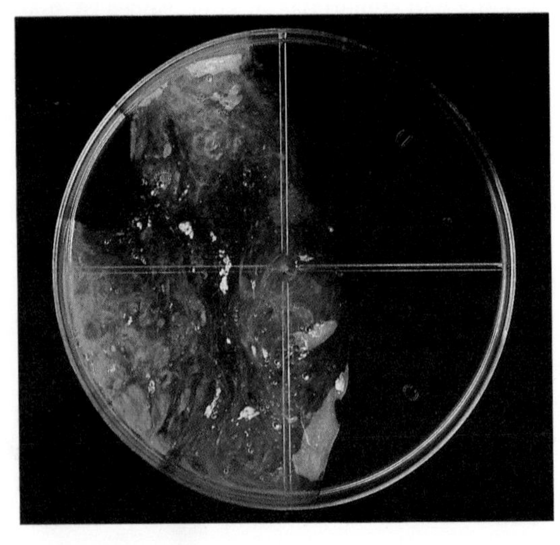

Lobar pneumonia blood in the sputum has a rusty red appearance and in pulmonary oedema much pink, frothy fluid is expectorated.

Sputum can be collected and submitted for micro-biological examination. The micro-biology examination provides information about:
- the type of microorganism;
- the pattern of antibiotic sensitivity, and therefore;
- the preferred choice of antibiotic.

Sputum can be submitted for cytological examination to determine the presence and type of malignant cells.

4. Summary

Coughing is a normal protective physiological phenomenon.
Abnormal increased coughing with or without production of mucus indicates a respiratory tract disorder. Many respiratory tract infections are spread by airborne droplets released into the air by coughing (and sneezing).
Sputum can be examined by bacteriological methods including culture and antibiotic sensitivity tests. Cytology examination for malignant cells can also be undertaken.

Tuberculosis is still, on a world basis, a common and important disease and often presents with haemoptysis.

Vomiting

28

1. Introduction

Most individuals have experienced episodes of vomiting, usually accompanied by nausea.
Although normally taken to indicate some underlying gastro-intestinal malfunction, vomiting occurs in many other diverse conditions. In some disorders, vomiting occurs without any co-existent nausea. Clinical history and examination usually give a sound indication of the underlying disorder.

Learning outcomes

After studying this chapter, the student should have sufficient knowledge and understanding of:
- the mechanism, cause, and effect of vomiting;
- the difference between vomiting and reflux;
- some side effects of vomiting;
- the risks of vomiting when conscious levels are lowered;
- the characteristics and causes of some different types of vomiting.

2. Manifestation of vomiting

In vomiting, the contents of the stomach, and sometimes of the duodenum, are forcefully ejected through the mouth by a vigorous gastric retrograde contraction. The gastric action is supplemented by abdominal muscle contraction. Vomiting is often preceded by a deep inspiration, and the epiglottis is positioned so as to prevent gastric contents entering the respiratory tract. Local stimulating nerve centres for vomiting are present in the pharynx, oesophagus, and stomach.

It is important to differentiate between *vomiting* and *reflux* or *regurgitation*. Although there is an anti-peristaltic action of the stomach and oesophagus in reflux or regurgitation, there is usually no nausea. The action of regurgitation is less precipitous than vomiting, and may be preceded by belching, heartburn, or epigastric pain. The deep inspiration and abdominal muscle contraction is absent in regurgitation. Nausea often precedes vomiting, but it is absent when the vomiting is triggered by a central nervous system lesion. Causes of vomiting in this state may include tumour, haemorrhage, or any lesion causing raised intra-cranial pressure. The vomiting thus caused may be so precipitous that it may be referred to as explosive. Pyloric stenosis in a young infant (usually male) causes characteristic projectile vomiting after a feed.

Patients with nausea and vomiting may suffer from numerous underlying conditions, including gastro-enteritis, gastric ulceration with or without haemorrhage, gastric perforation, peritonitis, pancreatitis and biliary disease. The side effects of many medicines may also cause nausea and vomiting.

3. Mechanism of vomiting

Immediately prior to vomiting the patient looks ill and pale, perspires, and shows marked salivation. During the act of vomiting, the pulse is slow due to vagal activity. These phenomena quickly disappear after vomiting ceases. The vomiting control centres of the central nervous system are partly in the medulla oblongata and partly in higher cortical centres, and both receive impulses from the local pharyngeal, oesophageal, or gastric centres. The presence of a so-called anti-vomiting centre is as yet controversial. It is probable that vomiting accompanying meningitis, brain tumour, encephalitis, or cerebrovascular accident arises by direct stimulation of the central centres, while causes arising in the gastro-intestinal system primarily stimulate the local centres, with subsequent transmission of nerve impulses to the brain.

Toxic material ingestion, whether by exogenous poison or drug side effect, may stimulate local or central centres or both. Medicines commonly causing vomiting include cytotoxics, antibiotics and digoxin. Metabolic disorders such as diabetes mellitus, liver failure, and renal failure commonly induce vomiting.

Vomiting in early pregnancy may be caused by central stimulation brought about by altered hormone and fluid balance levels.

4. Effects of vomiting

Prolonged and persistent vomiting leads to a serious upset of the fluid balance and to electrolyte loss. The developing dehydration and upset electrolyte balance themselves stimulate further vomiting. Adequate fluid and electrolyte replacement is necessary. In a patient weakened by persistent vomiting or in cases of lowered levels of consciousness, the respiratory tract is at risk from aspirated gastric contents as the normal protective mechanisms which close the epiglottis may fail. After a general anaesthetic or in a state of alcoholic intoxication the patient is particularly at risk. Great care must be exercised in any attempt to induce vomiting after intentional drug overdose as, again, conscious levels may be lowered and the airways compromised.

Prolonged retching and vomiting can lead to mucosal tears in the stomach and lower oesophagus leading to blood appearing in the vomitus. This is well recognised after alcoholic intoxication. Treatment of vomiting generally involves removal or neutralisation of any toxic irritant, correction of the fluid and salt balance, and, if necessary, protection of the airways together with administration of appropriate medication to lower local or central stimulus.

5. Types of vomiting

Continuous vomiting, despite the stomach being empty, can be referred to as dry vomiting or retching. Sometimes biliary material is ejected from the duodenum. This is seen in biliary colic, uraemia and peritonitis.
Where there are obstructive lesions at the pyloric end of the stomach (Figure 28.1), whether due to oedema, scarring or tumour, vomited material may contain food ingested many hours previously. The term *retention vomiting* is sometimes applied.

6. Composition of vomitus

Determining the composition of the vomitus is important if a correct diagnosis is to be arrived at. The remains of food long since eaten may indicate retention vomiting (see above). A careful history will soon confirm or refute this possibility.
Bile and traces of blood in the vomitus indicates prolonged and violent vomiting. Blood present in the stomach for any length of time is altered to a brown granular appearance ('coffee grounds').

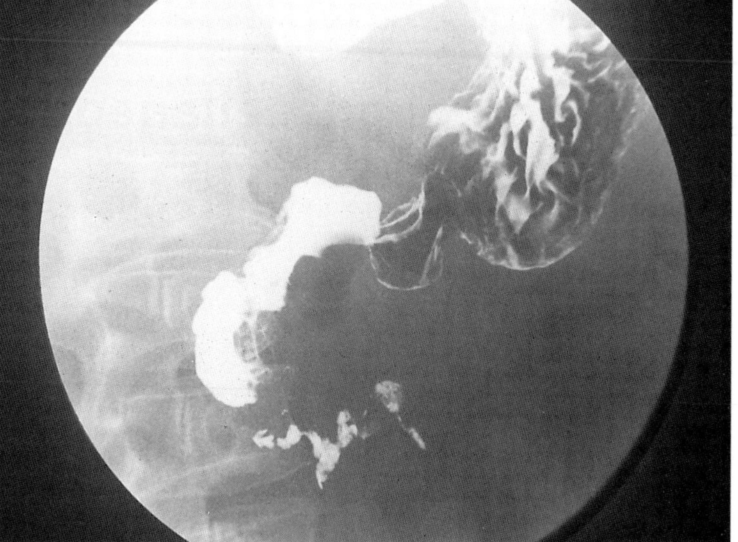

Figure 28.1
X-ray demonstrating duodenal obstruction

Fresh blood in the vomit represents recent haemorrhage. Vomiting of blood is called *haematemesis*, and when assessing the amount of blood loss in haematemesis, it must be remembered that much of the blood will have rapidly passed through the pylorus and will only show as melaena at a much later time.

7. Summary

Vomiting is most commonly caused by gastro-intestinal disorders. It may also occur, however, as a result of irritation of the vomiting centre in metabolic disease, poisoning, central nervous system space-occupying lesion or in inflammation.

Medications frequently cause nausea and vomiting.

29 Examination of miscellaneous fluids and tissues

1. Introduction

The previous chapters have outlined the examination of blood, urine, faeces, sputum, and vomitus. Fluids from other sources and cell aspirates from many sites can also be sampled, and a few more important examples are briefly described here.

Formal tissue biopsies for histological examination provide accurate tissue diagnosis in many conditions, but their consideration is beyond the remit of this text. Interested readers will find details in standard texts on histopathology.

Learning outcomes

After studying this chapter, the student should be able to understand and explain the diagnostic value of examining:
- pleural fluid;
- ascitic fluid;
- pericardial fluid;
- cerebrospinal fluid;
- synovial fluid.

The student should also be able to recognise the value and advantages of the following procedures:
- sternal marrow puncture;
- exfoliative cytology;
- fine needle aspiration.

2. Pleural fluid

In healthy individuals, minimal fluid is present in the pleural space where it acts as a lubricant and where, by surface tension effects, it helps maintain the pressure differential to keep the lungs inflated. Abnormal amounts of pleural fluid accumulate in conditions such as congestive cardiac failure, inflammatory and infective conditions such as pneumonia and in malignant disease. A large volume of pleural fluid may displace the underlying lungs to such an extent that it causes respiratory difficulties.

The fluid can be sampled by needle puncture, (Figure 29.1) during which care must be taken to prevent subsequent air entry and pneumothorax formation (Figure 29.1). Fluid low in protein is called a *transudate* and is common in congestive cardiac failure. Blood and leucocytes in the sample may indicate an inflammatory process such as pneumonia. Cytological examination may reveal the presence of malignant cells and may indicate the origin. Breast carcinoma and lung carcinoma frequently metastasise to the pleura resulting in a malignant pleural effusion.

Primary malignant pleural mesothelioma is often diagnosed by examination of samples from a malignant pleural effusion. Mesothelioma almost always occurs after previous occupational exposure to asbestos, even if the exposure was many years in the past. This is another example of where a good clinical history is almost diagnostic on its own.

In a *pleural biopsy* a small specimen of pleural tissue can be obtained by transcutaneous needle sampling. A sample can also be obtained by thoracoscopic examination (Figure 29.2) where the pleural cavity is examined through a fibreoptic flexible tube and a specimen of tissue can be removed.

3. Ascitic fluid

Ascites is the accumulation of excess fluid in the peritoneal cavity. The fluid may be low in protein, which indicates protein malnutrition, liver failure, or cardiac failure. Cirrhosis of the liver following previous hepatitis or as a manifestation of alcoholic liver disease commonly causes ascitic fluid to form. Nephrotic forms of renal failure may also produce ascitic fluid.

The volume of ascitic fluid may be so large as to obstruct diaphragmatic movement and cause respiratory difficulties (Figure 29.3).

Apart from providing a diagnostic specimen, removal of the fluid may temporarily relieve the patient's symptoms, although large volumes of fluid will contain significant amounts of protein and when this protein has been removed the underlying condition may be exacerbated, resulting in a rapid re-accumulation of the ascites. Intravenous albumin supplements are usually required in such circumstances. In severe cases the umbilicus will appear everted and the abdomen will be dull to percussion. Examination of the ascitic fluid is performed in a similar way to that of pleural fluid. The protein content can be assayed and samples submitted for microbiological study. Cytology examination will reveal any inflammatory or reactive cells, together with the presence of malignant cells. Common malignancies demonstrated in ascitic fluid include gastric carcinoma, colo-rectal carcinoma, pancreatic

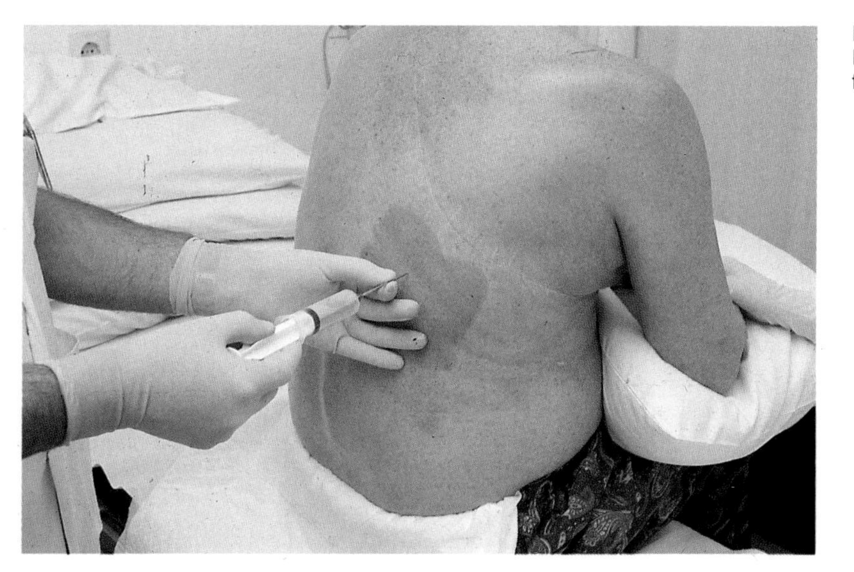

Figure 29.1
Removal of pleural fluid from the thorax

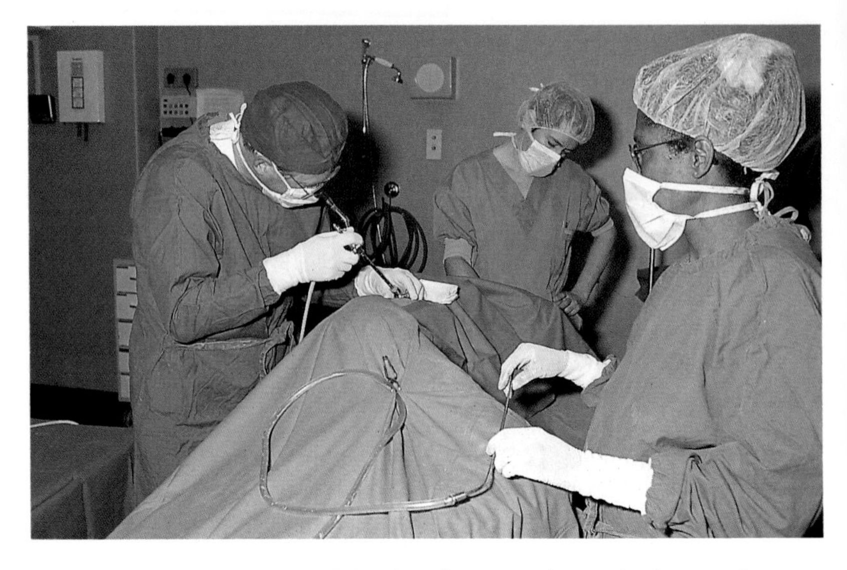
Figure 29.2
Thoracoscopy

carcinoma, female reproductive tract malignancy and tumours metastatic to the liver infiltrating the peritoneum and peritoneal space. Chronic accumulations of ascitic fluid often become loculated by fibrous bands.

4. Pericardial fluid

Removal of grossly excessive pericardial fluid is life saving, since it can mechanically impede the cardiac muscular contraction. Excess pericardial fluid forms in local inflammatory conditions (*pericarditis*). Spread from a nearby malignancy may lead to a malignant pericardial effusion. Examination follows the same techniques as seen above.

5. Cerebrospinal fluid

The cerebrospinal fluid surrounds the brain, is present in the cerebral ventricles and surrounds the spinal cord. It can best be sampled by *lumbar puncture* or *cisternal puncture* (Figure 29.4).

At the time of sampling the pressure can be measured. Fluid removed can be examined and cultured for evidence of meningitis or haemorrhage. Protein and enzyme assay may indicate multiple sclerosis and a variety of other neurological conditions. Viral studies may be diagnostic of encephalitis. Cytological preparations may show malignant cells originating from a primary brain neoplasm or from a cerebral metastasis.

Figure 29.3
Drainage of ascitic fluid (in a case of hepatic cirrhosis)

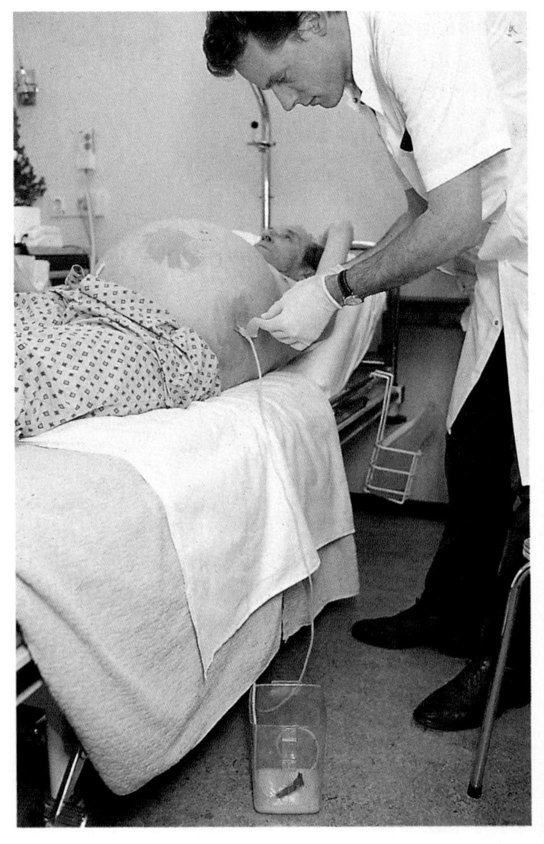

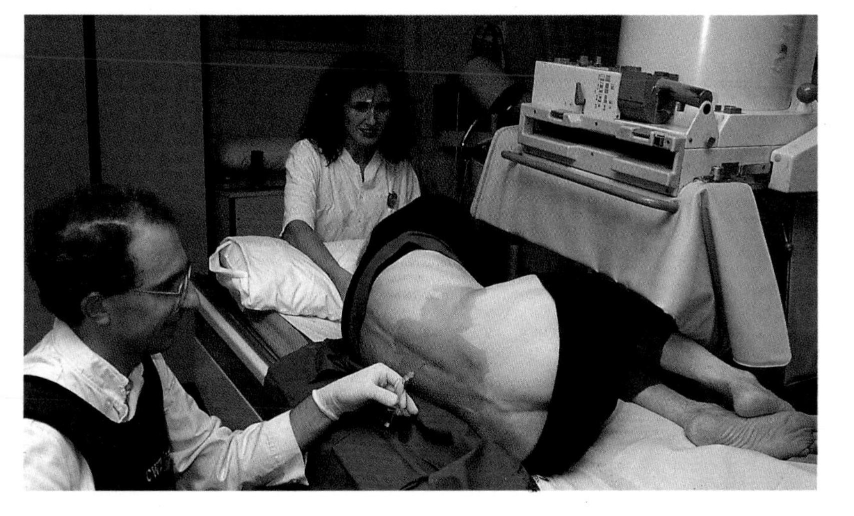

Figure 29.4
Lumbar puncture

6. Synovial fluid

Examination of synovial fluid may provide information to diagnose various joint swellings, including purulent arthritis, gout, and pyrophosphate arthropathy. Needle biopsy of the synovium can be undertaken by needle puncture.

7. Fine needle aspiration

Samples of liver and kidney are regularly obtained for histological examination by needle biopsy, and the use of fine needle aspiration for taking cytological specimens is increasing. Fine needle aspiration of the breast is frequently performed after suspicious mammography to establish the presence of breast carcinoma or benign breast disease. The limit of fine needle aspiration is determined only by the safe accessibility of the target site.

8. Marrow puncture

A needle inserted into the marrow cavity of the sternum or iliac crest will provide samples suitable for the diagnosis of many haematological conditions, including various anaemias, leukaemias and other malignant processes (Figure 29.5).

9. Exfoliative cytology

Cells can readily be obtained for examination by simply scraping them gently off their site of

Figure 29.5
Sternal puncture for bone marrow sampling

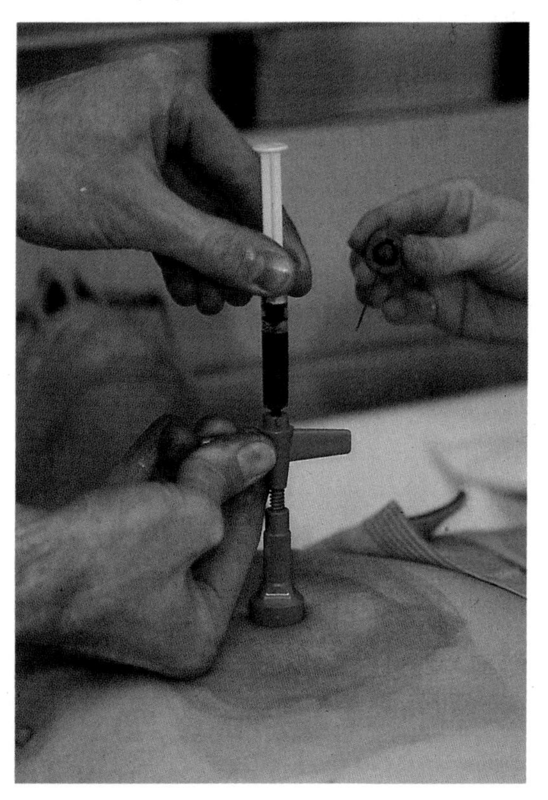

origin. Cervical cytology is the best known example. Suitably stained preparations can be assessed for hormonal status, presence of infection and for detection of dyskaryotic cells before invasive malignancy develops.

10. Summary

Clinical techniques now permit sampling of a variety of fluids accumulating in body cavities. Fine needle aspiration techniques are a cost-effective and relatively non-intrusive method of sampling many body sites for potential or suspected disease.

Appendix 1
Normal values

1. Blood count.
2. Coagulation studies.
3. Haemopoietic factors.
4. Biochemistry.
5. Blood gases.
6. Urine.
7. Thyroid function.
8. Adrenal function.
9. Pituitary/gonadal function.
10. Drugs and poisons.

BLOOD COUNT

	Adult Male	Adult Female
Haemoglobin	135 – 180	115 – 164 g/l
Red cell count	4.5 – 6.5	3.9 – 5.6 x 10 12/l
Packed cell volume	0.40 – 0.54	0.36 – 0.47
Mean cell volume	82 – 100 fl	
Mean cell haemoglobin	27 – 32pg	
Mean cell haemoglobin concentration	310 – 360 g/l	
White cell count	4.0 – 11.0 x 10^{9}/l	
Red cell size distribution width (RDW)	10.9 – 15.7	
Platelet count	150 – 400 x 10^{9}/l	
Plateletcrit (PCT)	0.15 – 0.32	
Mean platelet volume (MPV)	6.3 – 10.1 fl	
Platelet size distribution width (PDW)	15.5 – 17.5	
Lymphocytes (LYMPH)	20 – 45% (1.5 - 3.5 x 10^{9}/l)	
Reticulocyte count	0.2 – 2%	
E.S.R. (Westergren)	0 – 14	0 – 20mm

Differential White Cell Count

Neutrophils		40 – 75% (2.5 - 7.5 x 10^{9}/l)
Lymphocytes		20 – 45% (1.5 - 3.5 x 10^{9}/l)
Monocytes		2 – 10% (0.2 - 0.8 x 10^{9}/l)
Eosinophils		1 – 6% (0.040 - 0.44 x 10^{9}/l)
Basophils	less than	1 % (0 - 0.1 x 10^{9}/l)

COAGULATION TEST

Prothrombin time	(INR)	1 - 1.2
Prothrombin time (Therapeutic range)		See below
Thrombotest	(INR)	1.0 - 1.2
Thrombin Time		13 - 15 seconds
Partial thromboplastin time		Control time +6s (usually 24-30)
Bleeding time		2 - 9½ minutes
Fibrinogen		1.5 - 4.5 g/l
Fibrinogen degradation products		Less than 10 mg/l

PROTHROMBIN/THROMBOTEST

Suggested INR ranges in various conditions

Clinical State	INR
Prophylaxis of deep vein thrombosis including surgery on high risk patients	2.0 - 2.5
Hip surgery and fractured femur operations	2.0 - 3.0
Treatment of deep vein thrombosis Pulmonary embolism Systemic embolism Prevention of venous thrombo-embolism in myocardial infarction Mitral stenosis with embolism Transient ischaemic attacks Atrial fibrillation	2.0 - 3.0
Recurrent deep vein thrombosis and pulmonary embolism Arterial disease including myocardial infarction Mechanical prosthetic heart valves	3.0 - 4.5

HAEMOPOIETIC FACTORS

Serum

Iron	13 – 35 umol/l
Total iron binding capacity (TIBC)	45 – 70 umol/l
% Saturation	16 – 60%
Vitamin B_{12}	160 – 900 ng/l
Folate	1.6 – 12.6 ug/l

RBC

Folate	110 – 530 ug/l

Note
Reduced levels of serum iron are common in the absence of iron deficiency, particularly in hospital in-patients.

Infection, trauma, operations and chronic disease can all result in a low serum iron when body iron stores are adequate. Usually in such conditions the serum TIBC is normal or low.

The serum TIBC is usually elevated in iron deficiency but it is also raised in women taking oral contraceptives.

The percentage saturation of the TIBC gives a good although not infallible indication of iron status. A low level suggests iron deficiency.

In assessing the significance of levels of serum haemopoietic factors account should be taken of the full blood count and other haematology data. In particular a diagnosis of pernicious anaemia should not be made on the basis of a low serum vitamin B_{12} without other supporting evidence.

BIOCHEMISTRY

Serum / Plasma

Renal	Sodium	135 – 147 mmol/l
	Potassium	3.8 – 5.0 mmol/l
	Creatinine	60 – 120 umol/l
Protein	Total protein	55 – 80 g/l
	Albumin	35 – 55 g/l
LFT	Alkaline phosphatase (ALP)	20 – 120 i.u.
	Bilirubin	5 – 17 umol/l
	Asp. Trans	up to 35 i.u./l
Bone	Calcium	2.0 – 2.6 mmol/l
	Phosphate	0.8 – 1.4 mmol/l
	Alkaline phosphatase (ALP)	20 – 120 i.u.
Miscellaneous		
	Ammonia	24 – 48 umol/l
	Bicarbonate	24 – 34 mmol/l
	Chloride	95 – 105 mmol/l
	Magnesium	0.6 – 1.1 mmol/l
	Osmolality	275 – 290 mmol/kg water
	Urate	Male up to 0.45 mmol/l
		Female up to 0.39 mmol/l
	Urea	2.5 – 6.5 mmol/l
Carbohydrates		
	Glucose	3.4 – 5.5 mmol/l
	Fructosamine (non-diabetics)	1.8 – 2.8 mmol/l
Diabetics		
Satisfactory Control		2.8 – 3.2 mmol/l
Mediocre Control		3.2 – 3.7 mmol/l
Poor Control		3.7 – 4.5 mmol/l
Very Poor Control		> 4.5 mmol/l
Lactate		0.4 – 1.4 mmol/l

Lipids

- Currently recommended values

Cholesterol	Desirable	< 5.2 mmol/l
	Borderline	5.2 – 6.2 mmol/l
	High	> 6.2 mmol/l
	Triglycerides	< 2.2 mmol/l

Proteins (Serum)

Total protein		55 – 80 g/l
Albumin		35 – 55 g/l
Globulin		15 – 37 g/l
IgG	Adult level	5.2 – 16.1 g/l
IgA	Adult level	1.1 – 4.2 g/l
IgM	Adult level	0.5 – 1.7 g/l
IgE	Adult level up to	30 KU/l
Complement –	C3	0.7 – 1.6 g/l
	C4	0.2 – 0.5 g/l
Alpha 1 Antitrypsin		1.8 – 5.3 g/l
Haptoglobin		0.7 – 3.8 g/l (Hb binding)
C Reactive proteins		< 6 mg/l

Enzymes

Acid phosphatase (ACP)		
	(Total)	0 – 5 i.u.
	(prostatic)	0 – 1.6 i.u.
Aspartate transaminase (AST)		up to 35 i.u.
Amylase (AMS)		up to 120 i.u.
Creatine Kinase (CK)		up to 175 i.u.
Gamma Glutamyl Transpeptidase (GGT)		
	Male	up to 60 i.u.
	Female	up to 45 i.u.

BLOOD GASES

Arterial Values	
pH	7.37 – 7.42
pCO_2	4.5 – 6.1 kPa
Std. bicarbonate	21.3 – 24.4 mmol/1
Base excess	-2.3 to + 2.3 mmol/1
pO_2	12 – 15 kPa

URINE

(Values depend on 'Dietary' intake)

Osmolality (random)		250 – 1200 mmol/kg water
(24 hr)		400 – 600 mmol/kg water
Sodium		130 – 260 mmol/24 hr
Potassium		35 – 90 mmol/24 hr
Calcium		2.5 – 7.5 mmol/24 hr
Phosphate		15 – 50 mmol/24 hr
Creatinine		9000 – 17000 umol/24 hr
Urea		250 – 600 mmol/24 hr
Urate		3 – 12 mmol/24 hr
Oxalate	Males:	189 – 478 umol/24 hr
	Females:	266 – 511 umol/24 hr

C.S.F.

Glucose	3.3 – 4.4 mmol/l
Protein	0.15 – 0.40 g/1
Albumin	0.1 – 1.6 g/l
IgG	0.008 – 0.064 g/l
IgG/Albumin ratio	up to 0.27

THYROID FUNCTION - (Serum)

Thyroxine (T_4)	50 – 150 nmol/1
T_3	92 – 117%
Free Thyopac Index (FTI)	50 – 150
Tri-iodothyronine (T_3)	1.0 – 3.0 nmol/1
T.S.H.	0.4 – 8 mu/1
Free T_4	10 – 20 pmol/1

ADRENAL FUNCTION

PLASMA

Cortisol	(Midnight)	140 – 300 nmol/1
	(9.00a.m.)	140 – 500 nmol/1
	($\frac{1}{2}$hr Post Synacthen)	Rise of at least190 nmol/1

URINE

'Free' Cortisol -	Male	50 – 350 nmol/1
	Female	33 – 290 nmol/1
4 Hydroxy 3 Methoxy Mandelate (HMMA)		10 – 35 umol/24hr

Cortisol assays carried out by RIA technique. Result invalidated by Prednisolone which cross reacts in the assay.

PITUITARY / GONADAL FUNCTION

WOMEN	Follicular	Midcycle	Luteal	Post Menopausal	Units
L.H.	5 – 25	25 – 50	5 – 25	> 20	U/l
F.S.H	1 – 6	5 – 11	1 – 5	> 20	U/l
OESTRADIOL	70 – 800	170 – 900	100 – 550	>150	pmol/l

			Units
PROLACTIN		<500	mu/l
TESTOSTERONE		<3.5	nmol/l
PROGESTERONE	Pre-ovulatory	: 0.5 – 5.0	nmol/l
	Post	: 16 – 64	nmol/l

A mid-luteal phase progesterone level over 35 nmol/1 is consistent with ovulation.

MEN	All Adults	< 60 yrs	> 60 years	
L.H.		2 – 10	2 – 33	U/l
F.S.H.		2 – 8	2.5 – 23	U/l
OESTRADIOL	<150			pmol/l
PROLACTIN	<180			mu/l
TESTOSTERONE	14 – 40			nmol/l

DRUGS AND POISONS

DIGOXIN

Therapeutic range	0.8 – 2.0 ug/1	(1.0 – 2.5 nmol/1)
Toxic level	over 3.0 ug/1	(3.8 nmol/1)
Borderline toxic level	2.0 – 3.0 ug/1	(2.5 – 3.8 nmol/1)

Serum for digoxin assays must be collected at least 6 hours AFTER last dose.

LITHIUM

*Therapeutic range	0.6 – 1.5 mmol/1
Toxic level	over 2.0 mmol/1

*Serum for Lithium must be collected BEFORE morning dose.

THEOPHYLLINE

Neonates	7 – 15 mg/l	(ug/ml)
Adult/Child	10 – 20 mg/l	(ug/ml)

Type of preparation affects time taken to reach peak values -

slow release approximately 4 hours.
quick release approximately 2 hours.
Collect serum BEFORE oral dose.

LEAD (must be whole blood).

Adults and children	0.3 – 1.8 umol/l

Appendix 2
Specific methods of specimen collection

1. Nose, throat and pernasal swabs.
2. Sputum specimen.
3. Wound swabs, aspirates and surgical and biopsy specimens.
4. Eye swab.
5. High vaginal and cervical swabs.
6. Cervical swabs for chlamydial investigation.
7. Midstream specimen of urine.
8. Catheter specimen of urine.
9. Urine specimen for TB examination.
10. Faecal specimen.
11. Cutaneous specimens for fungal isolation.

Nose, throat and pernasal swabs

Aim: a) To isolate and identify agents of acute throat infection (bacteria and viruses).
b) To detect throat carriers of either Haemoltytic Streptococcus, Corynebacterium diphtheriae or pathogenic Neisseria spp.
c) To detect nasal carriers of the above mentioned.
d) To isolate Bordetella pertussis (whooping cough).

Nasal swab

Equipment: Sterile cotton-tipped swab stick
Stuarts transport medium
N/Saline 0.9% sachet
Waste container

Method: Moisten the swab with the N/Saline and gently rotate around the anterior nares.
Break off the swab stick into the transport medium.

Throat swab

Equipment: Sterile cotton-tipped swab stick
Stuarts transport medium and Virus transport medium (available from lab).
Spatula
Light source
Waste container

Method: Position patient comfortably and arrange for the throat to be properly illuminated.
Depress the patients tongue with the spatula.
Use the swab to obtain material from both fauces.
Transfer the swab into the required transport medium.

Pernasal swab

Equipment: A sterile swab at the end of a flexible piece of wire contained in a plugged test tube.

Method: Position the patient comfortably.
Introduce the pernasal swab below the inferior turbinate and into the post nasal space and gently rotate the swab to collect the sample.

Sputum specimen

Aim:
a) To provide a specimen of sputum from the lower respiratory tract.
b) To confirm or exclude pulmonary infection.
c) To assess the desirability of using chemotherapeutic agents in pulmonary infection.

Equipment:
Wide-necked, screw-capped sterile container
Tissues
Bag for waste disposal

Method:
Care should be taken to ensure the specimen sent for examination is sputum NOT SALIVA.
Instruct the patient to cough sputum directly into the container avoiding contamination with saliva and outside of the container.
The best time to obtain a specimen is in the morning when the patient awakes.
Encourage the patient to cough deeply. If problems obtaining an expectorated specimen are encountered, ask Physiotherapist for assistance.
If patient is suspected of having T.B., wear gloves when handling and replacing the lid on the container and any waste tissues.

Wound swabs, aspirates and surgical and biopsy specimens

Aim: a) The isolation and identification of pathogenic microorganisms responsible for the formation of abscesses and purulent discharges/infections.
b) The determination of antibiotic sensitivities of such organisms to enable effective treatment to be initiated or continued.

Equipment: Sterile cotton tipped swab stick
Sterile syringe and needle
Sterile screw-capped container or
Stuarts transport medium

Method: Always take a swab specimen before toiletry of lesion.
Place the sterile swab deep into the wound and collect specimen.
Break swab stick into transport medium.
Aspirated material may be placed directly into the screw-capped container; alternatively pus may be transferred directly to the container from a sterile receiver.
Do not place the container in contact with the lesion, to avoid contaminating the outside of the lesion.

Aspirates (eg. pleural, joint, peritoneal, pelvic, spinal fluids, etc) as well as surgical and biopsy specimens requiring microbiological examination are to be placed in sterile screw-capped containers with NO ADDED FIXATIVES OR PRESERVATIVES.

Eye swab

Aim: a) The isolation and identification of pathogenic microorganisms responsible for eye infections.
b) The determination of antibiotic sensitivities of such organisms to enable effective treatment to be initiated or continued.

Equipment: Cotton tipped swab stick
Transport media –
a) Stuarts
b) *Chlamydia* (available direct from lab).
c) Virus (available direct from lab).
Waste bag

Method: Eye swabbing should always be done before local anaesthetics are applied to the eye.
Hold the swab parallel to the eye and with your free hand part the patients eyelids, and gently rub the conjuncilva of the lower eyelid with the swab.
Transfer the swab into the appropriate transport medium.

NB Wipe off any purulent eye discharges before taking sample for chlamydia investigation.

High vaginal and cervical swabs

Aim: a) The isolation and identification of pathogenic microorganisms responsible for vaginal infections.
b) The determination of antibiotic sensitivities of such organisms to enable effective treatment to be initiated or continued.

Equipment: Vaginal speculum
Sterile cotton tipped swab sticks
Good light source
Sterile gauze swab
Container for waste disposal
Transport media –
a) Stuarts
b) Trichomonas
c) Chlamydia Virus (direct from lab).

Method: Put on gloves.
Place patient in dorsal position (lying on the back, knees flexed and widely separated).
With gloved hand part labia and introduce the speculum into the vagina and under good vision collect material from:–
a) posterior fornix
b) cervical os.
Carefully withdraw the swab, avoid contamination with lower vaginal and perineal flora.
Place the samples in the appropriate transport medium.

Cervical swabs for chlamydial investigation

Equipment: Vaginal speculum
Sterile cotton tipped swab sticks
Good light source
Sterile gauze swab
Container for waste disposal
Transport media –
Chlamydiazyme (direct from laboratory).

Method: Proceed as for high vaginal swab.
Then with sterile gauze or cotton wool balls wipe clean the cervix of mucopurulent discharge.
Introduce the swab into the cervical canal through the os and gently but firmly rotate the swab to collect material from the wall of the canal. Carefully withdraw after swabbing, avoid contamination with vaginal and perineal flora, and place in chlamydiazyme transport medium and replace cap.

Mid-stream specimen of urine

Aim: a) The isolation and identification of organisms infecting the urinary tract.
b) The determination of antibiotics or other chemotherapeutic sensitivity to assist in treatment.
c) To monitor response to treatment.
d) To detect urinary carriers of enteric organisms.

Equipment: Bedpan/toilet/urinal/receiver
Warm saline or soap and water for cleaning
Sterile gallipot
Cotton wool balls
Gloves
Specimen container
Receptacle for soiled disposables

Method: Wash hands and put on gloves.
In males: retract the foreskin and clean the glans penis and urethral meatus with soap and water/warm saline.
In females: clean the urethral meatus and vulva with soap and water/warm saline; swab with cotton wool balls from front to back.
Ask the patient to micturate into bedpan/toilet/urinal and then collect the midstream of the urine (20-30mls) into the specimen container (men) receiver (women).

Infants
Boys – After preliminary cleansing, a sterile tube is strapped to the penis to serve as a collecting vessel.
Girls – A polythene bag with adhesive fitting is placed over the previously cleansed perineum. (Alternatively a sterile latex glove may be used as a collecting vessel).

Catheter specimen of urine

Equipment: Alcohol swab
Sterile syringe and needle
Catheter clamp
Sterile specimen container
Receptacle for soiled disposables

Method: Clamp the drainage bag tubing below the rubber self-sealing port.
Wash your hands.
Cleanse the sleeve/port with alcohol swab.
Insert the assembled needle and syringe at a 45° angle into the rubber sleeve and 90° angle into the protected sampling port. (All sampling sleeves are designed to occlude puncture holes when needle is withdrawn).
Aspirate catheter urine into the syringe.
Withdraw needle and syringe from the catheter.
Re-swab the sampling sleeve/port with alcohol swab.
Transfer urine sample into container.
REMOVE CLAMP from the drainage bag tubing.

Urine specimen for T.B. examination

Collect three early morning urine specimens.

Faecal specimen

Aim: a) To establish the causation of diarrhoea.
b) The diagnosis of enteric levels.
c) The detection of faecal carriers of Salmonella and Shigelia.
d) The diagnosis of protozoal infestation.
e) The diagnosis of Staphylococcal enterocolitis.
f) The isolation and identification of enteropathogenic or toxigenic strains of Escherichia coti.
g) The diagnosis of pseudomembranous colitis.
h) The diagnosis of viral gastroenteritis.

Equipment: Bedpan
Spatula
Screw-capped container
Virus transport medium (available from the laboratory)
Toilet tissue
Receptacle for soiled disposables

Method: **Adults**: the stool is passed into the bedpan without urinating.
Using a spatula select (when possible) a suitable part with blood and mucus and place in container.
Wash your hands.
Infants: where possible a spatula should be used to collect material from soiled napkins and the specimen transferred to the container as above. (In other cases a rectal swab inserted at least $^1/_2$" above the anal margin may be used and the swab transferred into a Stuarts medium.)

Outpatients – Adults

The patient is issued with a screw-capped container and spatula and advised to proceed as follows.

a) Place in the bedpan a few loosely packed toilet papers.
b) Defecate onto the paper in the bedpan.
c) Collect a faecal sample from the paper using the spatula and transfer to the screw-capped container avoiding contaminating the outside of the container. Screw down the lid tightly, discard the spatula wrapped in paper/polythene bag into the household dustbin.
d) Wash hands with soap and water.

Outpatients – Infants

Instruct mothers to collect a specimen from the napkin as above. Alternatively a doctor or nurse may collect a rectal swab.
Samples for virological investigation should be put into virus transport medium (available from the laboratory).
Where AMOEBIC infection is suspected, please inform the laboratory and send a specimen without delay.

Cutaneous specimens (hair, skin, nail) for fungal isolation

Equipment: Alcohol wipe
Scalpel
Forceps
Gloves
Paper towel
Screw-capped container

Method: Wash your hands.
Put on gloves.

Hair – Remove distorted or fractured hairs with forceps.

Skin – Cleanse with alcohol wipe, before a specimen is collected, to reduce the level of contaminating skin flora.
Take shavings of epidermal scales at the active border of the lesion onto a paper towel using a scalpel.

Nail – Cleanse nail with alcohol wipe.
The outermost layer of the nail is then removed by scraping with a scalpel onto a paper towel.
Deeper scrapings, debris from under the edges of the infected nail, and nail clippings from infected areas are also suitable for culture.
All specimens should be placed in the screw-capped container.

References

Bailey, H. and Love, McN. (1981). *Short Practice of Surgery*, 18th edition, London, Lewis.

BMA and The Pharmaceutical Press. (1993). *British National Formulary*, No. 24, March 1993, London, BMA and The Pharmaceutical Press.

Gradwohl, R.B.H. (1976). *Legal Medicine*, 3rd edition, Bristol, John Wright.

Hoffbrand, A.V. and Lewis, S.M. (1972). *Haematology*, 1st edition, London, Heinemann.

Manson, P.F.C. (1987). *Tropical Diseases*, 19th edition, London, Bailliere Tindall.

Sternberg, S.S. and Raven, N.Y. (1989). *Diagnostic Surgical Pathology*, 2 vols.

Thomas, C.G.A. (1979). *Medical Microbiology*, 4th edition, London, Bailliere Tindall.

Zilva, J.F. and Pannall, P.R. (1984). *Clinical Chemistry in Diagnosis and Treatment*, 4th edition, London, Lloyd Luke.

The following are the latest available editions but not necessarily by the same publisher/title.

Bailey, H. and Love, McN. (1992). Revised by Mann, C.V. and Russell, R.C.G., *Short Practice of Surgery*, 21st edition, London, Chapman & Hall.

Ham, A.W. and Gleeson, T.S. (1961). *Histology*, 4th edition, Philadelphia, Lippincott.

Hoffbrand, A.V. and Lewis, S.M. (1981). *Post-Graduate Haematology*, 2nd edition, London, Heinemann Medical.

Zilva, J.F. and Pannall, P.R. and Mayne. (1991). *Clinical Chemistry in Diagnosis and Treatment*, 5th edition, reprint with correction, Sevenoaks, Kent.

Index